T0100641

Genetic Modification in the Food Industry

Genetic Modification in the Food Industry

A Strategy for Food Quality Improvement

Edited by

SIBEL ROLLER
South Bank University
London, UK

and

SUSAN HARLANDER
Vice-President
Pillsbury R&D
Minneapolis, USA

BLACKIE ACADEMIC & PROFESSIONAL
An Imprint of Chapman & Hall
London · Weinheim · New York · Tokyo · Melbourne · Madras

Published by Blackie Academic & Professional, an imprint of Thomson Science, 2–6 Boundary Row, London SE1 8HN, UK

Thomson Science, 2–6 Boundary Row, London SE1 8HN, UK

Thomson Science, 115 Fifth Avenue, New York, NY 10003, USA

Thomson Science, Suite 750, 400 Market Street, Philadelphia, PA 19106, USA

Thomson Science, Pappelallee 3, 69469 Weinheim, Germany

First edition 1998

Thomson Science is a division of International Thomson Publishing I(T)P®

Typeset in 10/12pt Times by Type Study, Scarborough, North Yorkshire

Printed in Great Britain by T.J. International Ltd, Padstow, Cornwall

ISBN 0 7514 0399 7

A catalogue record for this book is available from the British Library

Library of Congress Catalog Card Number: 97–75258

Printed on permanent acid-free text paper, manufactured in accordance with ANSI/NISO Z39.48–1992 (Permanence of Paper).

Contents

Contributors

Karen Barber
Campaigns Manager
Food and Drink Federation
6 Catherine Street
London WC2B 5JJ, UK

Paul Christou
Laboratory for Transgenic Technology
John Innes Institute
Norwich Research Park
Colney
Norwich NR4 7UH, UK

Lynn J. Frewer
Department of Consumer Studies
Institute of Food Research
Reading Laboratory
Earley Gate
Reading RG6 6BZ, UK

Peter W. Goodenough
School of Plant Sciences
University of Reading
Plant Science Laboratories
Whiteknights
PO Box 221
Reading RG6 6AS, UK

John Hammond
Head of Information and Communications
Brewing Research International
Lyttel Hall
Nutfield
Surrey RH1 4HY, UK

Susan Harlander
Vice President
Meals Division R&D and Agricultural Research
The Pillsbury Company
Pillsbury Technology Center
MS 9921, 330 University Avenue, SE
Minneapolis, MN 55414-2198, USA

Colin Hill
Microbiology Department
University College Cork
Western Road
Cork, Ireland

Anthony J. Kinney
DuPont Agricultural Products
DuPont Experimental Station
PO Box 80402
Wilmington, DE 19880-0402, USA

Paul Klaassen
Bakery Ingredients Divison
Gist-brocades
Wateringseweg 1
PO Box 1
2600 MA Delft, The Netherlands

Susan Knowlton
DuPont Agricultural Products
DuPont Experimental Station
PO Box 80402
Wilmington, DE 19880-0402, USA

Sibel Roller
Professor of Food Biotechnology
School of Applied Science
South Bank University
103 Borough Road
London SE1 0AA, UK

R. Paul Ross
National Dairy Products Research Centre
Moorepark
Fermoy
Ireland

Richard Shepherd
Department of Consumer Studies
Institute of Food Research
Reading Laboratory
Earley Gate
Reading RG6 6BZ, UK

David M. Stark
Director
NatureMark
A Unit of Monsanto Company
700 Chesterfield Parkway North
St Louis, MI 63198, USA

Roger Straughan
Department of Arts and Humanities in Education
University of Reading
Bulmershe Court
Earley
Reading RG8 1HY, UK

Nick Tomlinson
Head of Novel Foods and Processes Branch
Ministry of Agriculture, Fisheries and Food (MAFF)
Nobel House
17 Smith Square
London SW1P 3JR, UK

Rutger van Rooijen
Bakery Ingredients Division
Gist-brocades
Wateringseweg 1
PO Box 1
2600 MA Delft, The Netherlands

Preface

Several medical products of gene technology, such as human insulin from genetically modified microorganisms, have been available since the 1980s and have already made a tremendous beneficial impact on human health and welfare. In the food chain, however, developments in gene technology have been slower to emerge. Nevertheless, the first wave of research into the genetic transfer of herbicide, pest and disease resistance into food plants is now bearing fruit: in 1996, thousands of acres of genetically modified soybean, maize and rapeseed were planted in several countries. It has been predicted that the genetic modification of agronomic traits in plants will bring about a new 'Green Revolution' with benefits such as increased yields, reduced costs and reduced environmental damage accruing to the farmer and ultimately to the consumer.

The second wave of research in food-related gene technology concerns the modification of quality traits such as appearance, texture, flavour, shelf-life, ease of processing and nutritive value. Although technically more difficult to achieve, quality-improved foods are likely to bring even greater direct benefits to the food industry and the consumer than the agronomically improved crops mentioned above. The first genetically modified whole food with improved flavour (the tomato) appeared in the supermarkets in the USA in 1994. The true economic impact of this new technology on the food industry is yet to be realized.

As recombinant DNA technology is so new, many food scientists and technologists have little or no background in the science and have a limited concept of possible applications in foods. Yet, the products of gene technology will increasingly be on offer and decisions will need to be made as to whether or not to use them. The immediate aim of this book is to fill at least a part of the gap in knowledge among those working in the food chain so that they may, in turn, use the information to make scientifically based commercial decisions and to communicate with the public. The ultimate aim of this book is to foster a greater understanding of the benefits and potential pitfalls of gene technology as applied to foods.

As the ultimate success of food biotechnology will depend on consumer acceptance, Part 1 of this book consists of an in-depth discussion of fundamental societal issues including consumer perceptions, moral and ethical concerns, worldwide regulations and labelling, and communication with the public. A brief introduction to the basic techniques of genetic modification is also given in the introductory chapter to aid understanding of the case studies in Part 2.

In Part 2 of the book, groups of genetically modified foods or food ingredients that have recently become available or are to become available in the near future are reviewed in detail. In a book of this size, it would be impossible to cover all the genetically modified foods under development. Therefore, we have selected a limited number of genetically modified foods or food groups according to their likely commercial impact within the next 10–15 years. There are seven chapters on food enzymes, brewing and baking yeasts, starter cultures, oils, potatoes and cereals and they cover, whenever possible, issues such as history; production process; patent status; chemical and physical characteristics; performance in food processing operations; sensory aspects; health, safety and environmental issues; legislative and labelling position; consumer acceptance and marketing; and future prospects. The book ends with a glossary of technical terms.

Responsible scientists recognize that there are genuine concerns regarding the introduction of genetically modified organisms on the planet. The application of every new technology involves some risk and may produce unforeseen problems. The challenge will be to use the enhanced scientific tools and knowledge now available to attempt to predict problems and solve them before they happen. There is a need for public concerns to be addressed by continuous research and appropriate organizational and legislative measures.

Internationally known experts in governmental, industrial and academic organizations have contributed to this book. We would like to thank all the authors for their contributions, without whom a book of this nature could not have been written. The time and effort spent on the preparation of the chapters, and the authors' endeavours to accommodate our editorial requests, are much appreciated.*

Sibel Roller
Susan Harlander

*Views and opinions expressed by the authors of the various chapters are their own and do not necessarily reflect those of the editors.

Part 1
The Broader Issues

1 Modern food biotechnology: overview of key issues

SIBEL ROLLER and SUSAN HARLANDER

1.1 Introduction

Apart from occasional outbreaks of bacterial contamination, the consumption of food in the developed world is generally associated with relatively little risk to the health of the population. In recent decades, the focus of attention in industrialized societies has been on improving food quality, particularly as it relates to sensory and nutritional properties. The reduction of post-harvest deterioration of food materials, particularly fruit and vegetables, which can lead to massive economic losses, has constituted another important preoccupation of food providers in the industrialized world. Recent developments in recombinant DNA technology offer the opportunity not only of better controlling existing food processes, but also of developing entirely new approaches to quality control and food product development. The potential of this new technology to benefit the food processing industry and to improve food quality (e.g. appearance, texture, flavour, shelf-life, ease of processing and nutritive value) is enormous and its true economic impact has yet to be realized.

In terms of food quantity, it is worth remembering that little more than 100 years ago, famine was common in many parts of Europe and that this situation still arises in many tropical countries today. During the last decade, the levels of world food stocks have gravitated downwards from the FAO safety level of 60 days (Dunwell and Paul, 1990; IFST, 1996). Crop production remains vulnerable to climatic hazards, something which may be exacerbated in the future due to global warming. Furthermore, although the world's population has not reached the critical mass predicted in the 1960s, demographic trends may well precipitate food shortages in some countries. As has been pointed out by Dunwell and Paul (1990), in a country like the UK where only 2.1% of the population is employed in agriculture and where a population growth in the 1980s of 1.2% was accompanied by an increase of 11.4% in cereal production, it is easy to neglect developing countries such as Rwanda where, prior to the recent civil war, 92% of the population was employed in agriculture, and a population growth of 30.8% has been accompanied by a rise in cereal production of only 3%. Although some of these problems can only be remedied by political solutions, the new biotechnology offers an additional technological tool by which adequate food supplies could be assured for the entire world's population.

Therapeutic agents such as human insulin from genetically engineered microorganisms have been available for over a decade and have already had a tremendous impact on human health and welfare. By contrast, in the food industry, the products of gene technology are only just beginning to emerge. Of the genetically modified foods or food-related components developed to date, only recombinant chymosin has reportedly achieved substantial market penetration and is now used in the manufacture of as much as 70% of all cheese made in the USA (Guinee and Wilkinson, 1992; and Chapter 6). However, many other products of gene technology have recently received regulatory approval for use in foods, as shown in Table 1.1, reflecting an increase in the pace of development in food biotechnology. We can therefore expect to see many more genetically modified foods on supermarket shelves in the forthcoming decade. Ultimately, the commercial

Table 1.1 Commercial products of gene technology of interest to the food processing industry (excluding plants modified for agronomic traits)

Product name	Function	Producing company	Regulatory status and market penetration
Bovine somatotropin (BST)	Bovine hormone leading to increased milk yields	Monsanto, Eli Lilly	Permitted in 17 countries; used in USA since 1994; moratorium in EU until 1999, mainly for socio-political reasons
Chymosin	Replacement for calf rennet, supplies of which are inadequate	Gist-brocades, Genencor Intl., Chr. Hansens (Pfizer)	Permitted in Europe and USA; market penetration extensive in USA [see Chapter 6]
Baker's yeast	Faster carbon dioxide production, hence improved dough characteristics	Gist-brocades	UK-approved in 1990 but not currently used commercially [see Chapter 8]
Brewer's yeast	Able to degrade starch, hence useful for making reduced-calorie beer	BRF International	UK-approved in 1994; used on pilot scale [see Chapter 7]
Rapeseed oil (canola)	Increased content of lauric acid, therefore higher solid fat content suitable as replacement for chemically hydrogenated vegetable oil	Calgene/Monsanto	Contract grown since 1994 for personal care products (market penetration not known); food applications under development [more examples in Chapter 10]
Fresh tomato	Firmer texture, therefore can be allowed to ripen on vine, hence better taste	Calgene/Monsanto	First sold in USA in 1994; UK-approved in 1996; too early to determine market impact
Processing tomato	Higher solids, therefore cheaper to produce	Zeneca	First sold in UK in 1996; market impact not yet known

success of recombinant foods, like all new food products, will need to be demonstrated in the marketplace.

Currently, terms such as 'transformed', 'transgenic', 'recombinant', 'genetically engineered' and 'genetically modified' are all used to describe plants, animals or microorganisms which have had DNA introduced into them by means other than the combination of a sperm and an egg. The collective term 'genetically modified organisms' or GMOs is used frequently in regulatory documents and in the scientific literature. The term genetic modification, as opposed to 'engineering' or 'manipulation' is seen by some observers as a relatively neutral term with few emotive connotations and consequently is sometimes used in preference to the other expressions. In this book, as in most scientific and popular literature, we make no distinction between these different terms and use all of them interchangeably.

1.2 The basic techniques of genetic modification

The manipulation of genetic traits of plants and animals is not a new concept. Selective breeding has been practised for centuries, well before the molecular basis of the procedures was understood. All classical breeding relies upon and is limited by the inherent characteristics of a particular species. Thus, a black tulip cannot be produced by classical crossing because the gene coding for this characteristic is not inherent to this species of flower. The new recombinant DNA technology is different from traditional breeding methods in three principal ways: it reduces the random nature of classical breeding; it achieves the desired results much more quickly and predictably and, most importantly, it makes it possible to cross the species barrier for example, from an animal into a microorganism.

On a molecular level, the genomes of all living organisms work in a very similar way. Each living cell contains deoxyribonucleic acid (DNA) which carries the code for reproducing the entire organism. To use a familiar analogy, each gene can be thought of as a sequence of three-letter words, made from a four-letter alphabet, each letter being one of the four building blocks of DNA, that is the nucleotides adenine, cytosine, guanine and thymine. Each combination of three nucleotides is know as a codon. DNA codons are copied (transcribed) into messenger ribonucleic acid (mRNA) molecules which are, in turn, converted (translated) into strings of amino acids, that is protein, by the ribosomes. Each codon makes a single amino acid. Since there are 64 possible codons and only about 20 common amino acids in proteins, each amino acid can be made by more than one codon.

The transfer of genes from one cell to another can be likened to the old-fashioned editing process of 'cutting and pasting'. The genetic engineer uses very specific bacterial restriction enzymes that recognize, cut and ligate (join) DNA as molecular 'scissors and tape' to isolate the section of DNA coding for a specific trait (e.g. the ability to ferment a sugar). Since DNA does not readily move asexually from one organism to another, 'vehicles' known as plasmids are required to transfer the isolated gene from one organism into another. Plasmids are usually small rings of DNA that contain a limited number of genes. The new code is then inserted into the genome of the target organism so that it forms part of the normal cell machinery and is expressed together with all the other proteins that the organism normally produces. In addition to the new target gene, the tranformed organism may also contain regulatory DNA sequences such as promoters ('switch-on' signals) and terminators ('switch-off' signals) that control gene expression.

In plant biotechnology, a very specific plasmid from the plant pathogen *Agrobacterium tumefaciens* can be used, provided the tumour-inducing ability of the specific strain used has been excised (eliminated). The bacterium dissolves the plant cell wall and inserts DNA into the chromosome of the target cell. The entire plant can then be re-grown from the single modified cell. For more details on the *Agrobacterium*-mediated transfer of DNA in plants, see reviews by Zambryski (1992) and Hooykaas and Schilperoort (1992).

Since plasmids cannot be used in all applications, other methods of achieving DNA transfer have been developed. Ballistic bombardment or 'biolistics' (the use of small tungsten or gold particles coated with DNA and propelled into the target cell using gunpowder) was developed in 1990 and has been used successfully to modify several food crops, including rice and wheat (Klein *et al.*, 1992; and Chapter 12). Other methods of gene transfer involve electric pulsing or electroporation of pollen tubes, protoplasts (plant cells which have had their cell wall removed by enzymic treatment) or microorganisms. Newer methods, such as laser technology and micro-injection are also being developed.

DNA can now be sequenced and copied using automatic sequencers and synthesizers based on the polymerase chain reaction (PCR). The reaction has been likened to a 'genetic photocopier' and is based on two critical technological advances: the exploitation of the properties of thermostable bacterial DNA polymerase enzymes together with the development of automated thermal cyclers (PCR machines) (Graham, 1994). The most obvious application of PCR has been in detecting DNA in very small samples for diagnostic purposes as in genetic fingerprinting of criminal cases or in the detection of genetic disorders (Graham, 1994; Newton and Graham, 1994). Since its discovery in 1985, PCR has achieved widespread use in the research and clinical laboratory but remains to be taken up to any great extent in the food industry (see also section 1.8 on diagnostics).

Not all cells subjected to genetic modification techniques are successfully modified. Consequently, it is necessary to identify the modified cells (whether from plants, animals or microorganisms) using 'marker genes' closely linked to the genetic material to be transferred. This is often accomplished by using antibiotic resistance marker genes as, for example, in the case of the kanamycin-resistant marker gene used in the development of the genetically modified tomato. Such antibiotic resistance can be detected rapidly in the laboratory without the need to regenerate the whole plant or animal. In cereals, two frequently used markers are the phosphinothricin acetyl transferase (*pat*) gene or the closely related bialaphos resistance (*bar*) gene. These two genes originate from species of *Streptomyces* and enable the transformed organism to inactivate the normally toxic compounds phosphinothricin (an active ingredient of AgrEvo's herbicides *Basta*, *Challenge*, *Ignite*, *Finale* and *Harvest*) and bialaphos (an active ingredient of Meiji Seika's *Herbiace*), respectively (Dunwell, 1995; and Chapter 12).

The technique of cultivating undifferentiated plant cells (known as callus) *in vitro* and regenerating whole plants by adjusting the composition of the growth medium is worth mentioning here. This process, known as somaclonal variation, can be used to propagate many thousands of plants from just a few cells and to recover, at high frequency, the natural genetic variability from existing crop varieties as well as from genetically modified variants. Although somaclonal variation does not bring about genetic modification *per se*, many of the benefits of gene technology could not have been achieved without concomitant developments in plant propagation techniques (Evans and Sharp, 1986; Karp, 1993; Miller and Morrison, 1991).

1.3 Genetic modification of agronomic traits in crops

New crops modified for agronomic traits such as herbicide, insect and disease resistance have constituted the first wave of development in food-related biotechnology. By the mid-1990s, more than 80 crops had been genetically modified and field-tested (Table 12.2, Chapter 12) and more than 20 variants of maize, oilseed rape, soybean, potato and squash had received regulatory approval for large-scale sowing and use in foods in the USA and Canada (Tables 4.2 and 4.3, Chapter 4). By far the largest majority of the 10 000+ field trials carried out to date have been on traits of agricultural significance (Table 12.2C, Chapter 12).

Much of this work has been initiated and supported by the large agrochemical and seed companies because of the high initial cost of biotechnological development. Furthermore, it has been recognized that there is an increasing problem of resistance in some pest populations and there is a need to counteract the environmental consequences of intensive insecticide and pesticide use. In addition, the potential profitability of the market has also added impetus to R&D efforts; in 1996, the global market for agrochemicals for the top five crops grown in the USA and Europe (maize, cotton, wheat, oilseed rape and sugar beet) was estimated at over $9.5 billion (Lloyd-Evans and Barfoot, 1996).

Many agronomic traits in plants are relatively simple technical targets for genetic modification because they are determined by a single biochemical pathway which is, in turn, controlled by a single gene. For example, sensitivity to the herbicide glyphosate is controlled by the single enzyme 5-enolpyruvylshikimate-3-phosphate (EPSP) synthase. Glyphosate-tolerant soya, cotton and oilseed rape are already on the market in the USA and Canada (Chapter 4). Since conventional crops as well as weeds are killed by this herbicide, the use of glyphosate in conjunction with the GM variant allows for reduced levels of application, thereby conferring benefits for the environment (for a discussion of potential environmental impacts, see section 1.4).

The soil-living bacterium *Bacillus thuringiensis* produces specific proteins (known collectively as B.t.) active against pests such as beetles, moths, flies and worms. Several thousand tonnes of these protein preparations have been in use for a number of years as biopesticides with no detrimental effects on the environment. However, because of their poor stability and expense, B.t.-based biopesticides have not achieved extensive market penetration. To get around this problem, several genes encoding for insecticidal B.t. proteins have now been inserted into plants including cabbage, soyabeans, maize, potatoes and rice. By the mid-1990s, several variants of transgenic maize producing B.t. against the Euorpean corn borer had been developed and, in 1996, Ciba Geigy's GM maize was the first cereal to be planted commercially after receiving regulatory clearance in the European Union, the USA, Canada and Japan (Chapters 4 and 12; Koziel *et al.*, 1993).

Ultimately, it is conceivable that crops with a combination of recombinant genes could be tailored to suit a variety of specific localities and/or climatic conditions to give optimum agronomic performance. However, the development of such 'designer crops' may well be some way in the future.

1.4 Environmental concerns

The development of herbicide-resistant crops has provoked much controversy, particularly in Europe, with proponents predicting reduced reliance on herbicides and critics fearing a potential increase in the use of these important agrochemicals (Rissler and Mellon, 1994; van Wagner, 1993). As might be expected, the true picture is rather more complex.

Current agronomic practices involve the preventive application of several selective herbicides before weed infestation has started (Dunwell, 1994). Given the uncertain effectiveness of some older herbicides, the farmer may often apply these 'pre-emergent' herbicides in excess of requirements, with potential detrimental consequences for the environment. It has been argued that the new genetically modified crops will allow farmers to use newer, more ecologically friendly, broad-spectrum herbicides at lower but more effective concentrations and only when weed pressure actually demands it. The availability of the new herbicides coupled with 'matching' genetically modified crops could change agronomic practices profoundly, with the environment as the ultimate beneficiary.

One of the concerns frequently expressed about herbicide-resistant plants is their potential for becoming weeds that can stifle useful crops and wild plants. The characteristics of weeds are usually controlled by many complex genes and it is highly unlikely that the genetic manipulation of a single gene trait in a crop would increase its tendency to become a weed (Taylor and Thomas, 1994). Of greater concern is the risk of the herbicide-resistant gene spreading through pollen to plants which are already weeds. Although field studies have shown that pollen from genetically modified plants is rarely carried over very long distances, the possibility nevertheless exists that this may occur (there are historical examples of genes moving from one organism to another by unknown mechanisms over several million years) and may eventually result in the transfer of the genetically modified trait to a weed. However, these occurrences are so rare that it would be impossible to monitor them in practice (Taylor and Thomas, 1994).

When considering the potential threat of a genetically modified plant to the environment, several properties of the crop must be taken into account. For example, maize plants have no seed dispersal mechanism as the seeds are wrapped in cobs and adhere tightly to each other (Taylor and Thomas, 1994). Therefore, gene transfer can be expected to be limited as there are no wild relatives of maize in the USA, where much of the global crop is planted. However, in Mexico and Central America, where maize originated and where wild relatives exist, careful monitoring may be necessary (Morrow, 1996). Similarly, wheat has no wild relatives in the USA. However, other plants may cause more concern. For example, the cultivated oat and sorghum cross readily with their wild relatives many of which are serious agricultural weeds (Morrow, 1996; Taylor and Thomas, 1994); therefore, genetic modification for multiple herbicide resistance in these crops may well pose a threat to the environment. Clearly, each genetic modification and application to a specific crop needs to be scrutinized on a case-by-case basis. These and other environmental concerns have been reflected in the regulations, as discussed in Chapter 4 (FDA, 1992; Olempska-Beer *et al.*, 1993).

The increasing number of field trials of GM plants conducted in the last decade without any adverse environmental effects represents considerable accumulated experience of benefit to developers and regulators alike. Furthermore, there is a growing body of evidence arising from ecological studies which suggests that some of the concerns regarding potential environmental damage may have been over-emphasized (Crawley *et al.*, 1993).

1.5 The genetic modification of quality in foods

While the genetic modification of agronomic traits in crops may bring about dramatic changes in the supply and quantity of food available before the turn of the century, improvements in food quality brought about by the genetic route are likely to develop more slowly for both technical and regulatory reasons. Nevertheless, the first example of a quality-improved whole food, the GM tomato, is already on the supermarket shelves in the USA.

Alterations in the macromolecular components (protein, carbohydrate and lipid) of certain foods are the most obvious quality targets for genetic engineers. For example, the addition of extra genes coding for the high molecular weight subunits of gluten in wheat has been attempted as a means of improving the dough-making quality of certain soft wheat flours (Hassler, 1995; Shewry *et al.*, 1995). As another example, a bacterial gene has been inserted successfully into potato plants to increase the overall proportion of starch in the tubers while reducing water content, resulting in reduced absorption of fat during cooking in oil (Chapter 11). This development could be of great interest to the health-conscious consumer who is concerned about excessive intake of fat. Furthermore, the alteration of the amylose/amylopectin ratio in starch-bearing crops including potatoes and maize has been reported (Smith and Martin, 1994; Visser and Jacobsen, 1993; Dunwell, 1994).

Of the macromolecular modifications to date, those involving lipid composition may be nearest to commercial realization. For example, rapeseed oil (canola) has been modified genetically to contain up to 40% lauric acid which in turn gives the oil a higher solid fat content (Chapter 10). The recombinant rapeseed was contract-grown in Canada in 1996 with primary commercial targets in the soap and cosmetics industries. It has been argued by some industrial observers that the inevitable higher cost of producing higher value oils from genetically modified plants may not be sustainable due to the commodity nature of the oilseed crop. Nevertheless, potential applications for high-laurate oil as a cocoa butter substitute have been suggested. Another modified plant oil very close to commercialization is the high oleic acid soybean, described in more detail in Chapter 10. It is conceivable that in the near future, oils high in mono- and polyunsaturated fatty acids may be developed for use in 'healthy' food applications whereas high-stearate oils may replace hydrogenated fats in processed foods.

Many plants produce antinutritional factors which may cause poisoning or even death if ingested in large quantities. For example, many legumes contain lectins which, if not removed by soaking and cooking, can cause severe gastrointestinal symptoms. Genetic modification has been used to produce plants low in these antinutritional factors; however, because these compounds play an important role in the plant's resistance to pests and pathogenic microorganisms, the GM variants frequently suffer from increased sensitivity to plant diseases and consequently yield poorer harvests. Clearly, further advances in both fundamental and applied knowledge will be necessary before the practical products of these efforts can enter the markets.

Developments in the genetic modification of compositional and sensorial traits of plants have been relatively slow due to a lack of fundamental knowledge about the complex combinations of genes controlling these qualities. This gap in knowledge is being addressed by a number of large genome sequencing projects that have been undertaken, usually on a collaborative basis, across Europe and the USA. For example, the weed *Arabidopsis thaliana* (a member of the cabbage family) is being studied as a relatively simple model system, from which it will be possible to draw conclusions relevant to the important, but more complicated, industrial crops. Genome databanks on many food plants including soya, rice, maize, wheat, barley, rye, sorghum, the solanaceae (tomato, etc.), beans and peas are being developed progressively as more and more information about the gene sequences becomes available (Hansen and Magnien, 1994). Once these detailed genome databanks are available, further advances in quality improvements in foods should be more rapid than has hitherto been possible.

1.6 The genetically modified tomato

The genetically modified tomato (Calgene's fresh Flavr Savr™ and Zeneca's processing tomato) represents one of the success stories of modern food biotechnology and as such cannot be omitted from any review of the subject. However, much has been written about the scientific, marketing and regulatory development of the GM tomato and consequently, details will not be repeated here. The interested reader is referred to the many references given below, including a book on the regulatory assessment of Calgene's tomato by the FDA (Redenbaugh *et al.*, 1992), a subsequent review paper by Calgene authors (Redenbaugh *et al.*, 1994) and a review of tomato ripening (Hobson and Grierson, 1993).

1.6.1 The need for a tomato with improved flavour

Although most large conurban supermarkets now stock up to half a dozen or more different varieties of fresh tomatoes on a year-around basis, it is generally agreed that, in terms of flavour, the modern, mass-produced tomato is a poor relative of the traditional product grown seasonally 'in the back yard'. This situation has arisen because of the need for the modern tomato to be picked green and firm in order to withstand the rigours of mechanical harvesting and long-haul transport. Following the long journey from farm to supermarket, the immature tomato is often treated with ethylene gas to produce the bright red colour associated with ripe fruit. Unfortunately, this treatment fails to induce the production of the characteristic flavours of tomato. The poor sensory qualities, combined with the relatively easy technical target (tomatoes have a small genome and can be grown easily in tissue culture), and the very large markets for fresh (estimated at $5 billion [Pfeiffer, 1994] in the USA alone) and processed tomatoes, have led several companies to develop genetically modified variants with improved properties.

1.6.2 The inhibition of pectin degradation in the GM tomato

During ripening of tomatoes, the enzyme polygalacturonase degrades the polymeric backbone of pectin resulting in softening of the fruit (Hobson and Grierson, 1993). Therefore, the main goal of tomato engineering has been to eliminate or delay the action of this enzyme in order to retard or prevent softening. In this way, the GM tomato can be allowed to remain on the vine to ripen and develop flavour to its full potential without losing firmness and becoming too soft for transport. Thus, superior taste is achieved indirectly, that is without modification of the more complex biochemical pathways involved in flavour production but simply by making it possible to delay premature harvest. In addition, the added bonus of extended shelf-life at retail level is achieved.

The inhibition of polygalacturonase activity in transgenic tomatoes has been achieved by either reversing the orientation of the DNA coding for the enzyme (known as antisense technology and used by Calgene) or by truncating the DNA encoding for the enzyme (Zeneca's method). The end-effect of the two approaches has been the same: enzyme activity is modulated by reducing the expression of the protein (Schuch, 1994). The inhibition of the enzyme has been achieved without interfering with other aspects of the tomato's natural ripening mechanism including colour and flavour development.

The overall quality and acceptability of many tomato products including sauces, pastes, ketchup, soups and juices, is determined by the ultimate viscosity and consistency of the product of which undenatured pectin is the most important determinant. Most tomato processing operations begin by the preparation of tomato pulp. During manufacture of tomato pulp destined for products in which the thickening properties of pectin are to be retained as far as possible (e.g. tomato paste), the 'hot-break' procedure is used. In this process, fresh tomatoes are heated at 96°C for 5–10 minutes to destroy the pectic enzymes. Unfortunately, this process, as well as being energy intensive, also results in some loss of flavour volatiles from the product. The lack of polygalacturonase activity in the genetically modified processing tomato means that the hot-break process can be eliminated or, rather, the temperatures used during comminution can be reduced. Thus, the new processing tomato is bringing benefits for the food processor (reduced energy costs), the environment (reduced waste) and the consumer (improved flavour and reduced cost of pastes).

1.6.3 Performance, safety and approval of the GM tomato

The achievement of regulatory approval for the transgenic tomato has been a lengthy process probably because the product was the first whole food to be genetically modified for a quality parameter and the regulatory authorities, as well as the industrial applicants, were both following a steep learning curve about the products of biotechnology. To obtain approval, Calgene has submitted to the FDA results of comprehensive studies on the composition and safety of their Flavr Savr™ tomato, carried out over a period of eight years (Redenbaugh *et al.*, 1992; Anon., 1993; Miflin, 1993). Detailed analyses of the protein, vitamins A, B and C, niacin, calcium, magnesium, phosphorus, sodium and iron contents, as well as of the natural toxin tomatine, were undertaken (Redenbaugh *et al.*, 1992). Technological performance of the recombinant tomato (e.g. acidity, viscosity, consistency and pH in juice and pastes) was determined, and acute toxicity trials in laboratory animals were carried out (Kramer *et al.*, 1990, 1992; Redenbaugh *et al.*, 1992). The results of more than four years of field trials in Florida, California and Mexico were also made available to the FDA. Interestingly, recombinant fruit was inadvertently found to be more resistant to certain common fungal infections during the course of these studies (Kramer *et al.*, 1992). Apart from the difference in viscosity arising from the presence of undegraded pectin in the ripe fruit, the recombinant tomato was essentially the same in composition as the wild-type fruit.

In 1994, the Flavr Savr™ tomato received clearance from the FDA and by the end of May 1995, Calgene was supplying it to over 1700 retail outlets in selected states in the Mid-West, New England and the West Coast (Anon., 1995).

Similarly, in the UK, Zeneca's processing tomato received regulatory clearance in 1995 and the resultant paste was sold in selected supermarkets in 1996. In the same year, the recombinant paste received clearance from the FDA. Zeneca had collaborated with the Peto Seed Company since 1989 on this development although this was preceded by many years of fundamental work in tomato biochemistry by Professor Don Grierson's team at Nottingham University in the UK. In early work, trangenic tomatoes were tested under conditions designed to prevent any release of the experimental material to the environment. In line with the UK's Environmental Protection Act, the tomatoes were grown in sealed greenhouses to prevent pollen escaping and all seeds were removed from the fruit prior to tasting trials to avoid the risk of release into sewers (Anon., 1991). Field trials were carried out in Chile and California and technological performance tests showed that the tomato lines in which polygalacturonase activity had been reduced by 99% yielded pastes with a viscosity that was 80% greater than in conventional lines (Schuch, 1989).

The parallel developments at Calgene and Zeneca have led to several bitter patent disputes over the technology used to achieve the end result. However, in 1994, the two companies ended the dispute with an agreement that Calgene would own exclusive world rights to fresh GM tomatoes whilst Zeneca would hold the rights to use the tomatoes in processed foods including soups and ketchup (Anon., 1994).

1.6.4 The marketing of the GM tomato

The introduction of the GM tomato to the food markets has been a prime example of a successful communication strategy between industry and the consumer. For example, in the UK, labelling of the GM tomato was not required by law and yet the two retailers offering the GM paste in their supermarkets in 1996 had decided to voluntarily label the products. Sainsbury's supermarkets labelled their paste as 'Made with genetically modified tomatoes' on the front of the can; in addition, the side of the can was labelled with the statement: 'The benefits of using genetically modified tomatoes for this product are less waste and reduced energy in processing.' To cap it all, the new tomato paste was priced at about 10% less than conventional paste thereby offering the consumer an obvious financial benefit. The Safeway tomato paste has been similarly labelled. Both supermarket chains have also produced explanatory leaflets available from their customer information desks for interested consumers (Anon., 1996).

1.6.5 Prospects for quality improvements in fruit and vegetables

Many other genes responsible for controlling other quality aspects in tomatoes such as lycopene (red colouring), soluble solids (sugar content), the sugar:acid balance, volatiles, etc., have been identified (Bennett *et al.*, 1995; Schuch, 1994). Antisense technology has also been used to control the texture of other fruit and vegetables. For example, GM broccoli, raspberries, and melon with improved ability to withstand the rigours of long-haul transport have been reported but none of these has entered the markets as yet (Bleecker, 1989).

1.7 Starter cultures for fermented foods

The use of microorganisms in the food industry has a long history and fermented foods such as yoghurt, cheese, salami-type sausages, bread, beer and wine, sauerkraut and soy sauce, tea and coffee, etc. are well-accepted by consumers. Newer microbiological products have also been successful. For example, 'bio' yoghurts have become very popular due to claimed digestive benefits resulting from the particular live organisms they contain. Modern genetic methods now make the development of improved starter cultures for fermented foods possible in a fraction of the time needed using conventional techniques. Some of the desirable traits which have been the focus of genetic investigations include flavour production, resistance to viruses (bacteriophages) and bacteriocin (antimicrobial) production and are discussed in more detail in Chapter 9.

In 1996, the sequencing of the genome of the very important food yeast *Saccharomyces cerevisiae* was completed within a worldwide network of laboratories from the USA, Europe and Japan (Goffeau, 1996). Current work is focusing on identifying the function of the individual genes. Once this has been completed, many more applications in food as well as other industries should be more readily accessible. Meanwhile, two very useful genetically modified strains of brewer's and baker's yeasts have been developed and are discussed in more detail in Chapters 7 and 8, respectively.

1.8 Diagnostics for food testing

Legislative and consumer demands for foods of consistently high quality and guaranteed safety have led to increasing pressure on industry to develop simple, reliable, rapid and cost-effective analytical techniques. Many conventional methods of testing foods, particularly in microbiology, are increasingly being regarded as unacceptably long, cumbersome and not wholly reliable. Together with many other biologically-based tests, DNA-based diagnostics are beginning to make an impact on the food industry due to their exceptional specificity and very high potential sensitivity and rapidity. For details on developments in the diagnostics arena, the interested reader is referred to a review and book by Patel (1994a,b).

Gene probes can be tailored to detect microorganisms at the genus, species or subspecies level. The probes are generally constructed to contain complementary bases which hybridize with sections of DNA (or sometimes RNA) specific to the target microorganism. Hybridization is revealed by tagging the probe with an identifiable marker which gives a detectable signal, for example, fluorescence (Swaminathan and Feng, 1994).

Commercial probes for *Listeria*, *Salmonella*, enteropathogenic *Escherichia coli* and *Campylobacter jejuni/coli* first became available in the mid-1980s although these early products had a relatively low impact in the food industry. A major problem was the relatively high number of organisms (approximately 10^5 cells/ml or more) needed for detection by the early DNA probes. Consequently, many pathogenic organisms which may be present at very low levels in a food needed to be enriched using conventional cultural methods thereby adding about two days to the detection time. Recently, it has become possible to avoid this problem by amplifying nanogram quantities of DNA using PCR (polymerase chain reaction). This technique has been commercialized to detect *Legionella* at a level of 100 cells/ml in 4 h. Similar developments in the detection of low levels of the food-borne pathogen *Salmonella* have also been commercialized (Graham, 1994; Patel, 1994a; Swaminathan and Feng, 1994).

Nucleic acid probes with adequate robustness and sensitivity require extensive development before being available sufficiently cheaply to attract users in the food industry. Presentation of gene probes in card formats has been developed for clinical uses and may well be developed for food applications in the future (Graham, 1994; Swaminathan and Feng, 1994). However, the cost constraints as well as various sample collection and preparation problems may well result in a delay in adoption.

DNA probing has already been used in many research applications and more recently in clinical testing for human genetic diseases and in the forensic laboratory. It is anticipated that DNA diagnostic testing particularly based on PCR (polymerase chain reaction) amplification will also become an important part of food testing during the next decade. Although the greatest impact is likely to be in food microbiology, other applications such as in authenticity testing (e.g. meat speciation and detection of specific citrus cultivars in fruit juice blends) are also conceivable. Unlike the development of foods and food ingredients destined for human consumption, the development of DNA-based diagnostics is less likely to be constrained by regulatory hurdles.

1.9 The genetic modification of animals

The genetic modification of animals has, to date, presented a technical challenge primarily because novel genes have to be introduced at a very early stage of development if all the cells in the mature animal are to contain them. Microinjection of DNA into newly fertilized eggs followed by transfer into the uterus of a foster mother is a highly inefficient process although it has met with some success in cows, pigs, sheep and goats.

Many of the potential applications of biotechnology in animals are currently not possible due to limited knowledge about the structure and function of animal genes. Consequently, several genome mapping projects are well underway in Europe and the USA. For example, the entire genomes of the pig (PigMaP), the cow (BovMaP) and the domestic hen are being sequenced with the aim of ultimately developing new diagnostic assays, vaccines and therapeutic agents for the animals (Anon., 1996; Hansen and Magnien, 1994).

Most work on the genetic modification of livestock has focused on the production of valuable pharmaceutical proteins in milk. Although transgenic animals carrying extra copies of growth hormone genes have been produced and have been shown to produce leaner meat more efficiently, the general health and reproductory status of the animals has suffered and consequently many similar experiments have been stopped (Houdebine, 1997). Ethical objections to animal work are likely to represent a major obstacle to further work in the development of improved meat quality using transgenic techniques. It will be interesting to see whether similar ethical concerns are expressed when the first genetically modified fish, such as the faster-growing salmon that is currently under development, come up for safety assessment and regulatory approval (Chen, 1994).

1.9.1 Herman the transgenic bull

A notorious example of a potential application arising from fundamental genome projects has been Herman the transgenic bull. Herman was produced in The Netherlands in 1992 when an extra copy of the gene encoding for lactoferrin (an antimicrobial protein naturally present in milk) was injected into several embryos thereby conferring increased resistance to infection by mastitis. Unfortunately, the only embryo that implanted successfully was male and consequently Herman had to produce female offspring before the success of the gene transfer could be ascertained. The case attracted much publicity because permission to mate Herman had to be obtained from the Dutch parliament.

1.9.2 Dolly the cloned sheep

On 22 February 1997, scientists from the Roslin Institute and PPL Therapeutics in Scotland attracted global news coverage by announcing the creation of the first cloned mammal, Dolly. Dolly was produced by replacing the genes in a normal sheep egg with DNA extracted from an adult sheep's mammary gland and returning the construct into the uterus of a foster mother for full gestation. The first successful product of several hundred abortive attempts at the technique, Dolly provoked an intense public reaction (both positive and negative). The ensuing debate about ethics, morals and the safety of animal biotechnology highlighted the difficulties of working with animals which are perceived as sentient (for further discussion on the ethical issues, see Chapter 3).

1.9.3 Recombinant BST

Although the hormone bovine somatotropin (BST) from GM bacteria is not a food or food ingredient *per se*, its development and introduction into animal husbandry practices in some countries is of relevance here. The hormone is normally present in cows and can be found in low concentrations in all cows' milk. BST plays a role in the regulation of milk yield, growth rate and protein-to-fat ratios and is used commercially mainly to increase milk yields. Since the supply of BST from the pituitary glands of slaughtered cows was limiting the usefulness of BST in agriculture, scientists at Monsanto inserted the gene coding for the hormone into *Escherichia coli* thereby allowing for large-scale production by fermentation. Eli Lilly, Upjohn and Dow Elanco have also been active in the area.

In the early stages of the BST approval process in the USA, several vociferous anti-biotech groups succeeded in launching a boycott of food containing milk from BST-treated cows and received the backing of several food companies including an ice-cream maker and a manufacturer of infant formulae (Fox, 1993b). Although the issue appears to have died a death in the USA, in Europe the introduction of recombinant BST into the marketplace continues to be controversial. Concerns have centred mainly on four issues: human health, animal health and welfare, labelling, and socio-economic aspects.

Extensive scientific investigations (the FDA and Monsanto have compiled some 500 volumes of approximately 400 pages each of compositional and safety data on recombinant BST) and 3 years of practical experience of actual use in the USA have shown that many of the original fears expressed about BST were groundless (Bauman, 1992; Bauman and Vernon, 1993). Nevertheless, in the European Union, recombinant BST has met with fierce political opposition and a moratorium on its use has been imposed until 1999 mainly because of fears of a negative effect on the social and economic welfare of the European farmer.

In view of the high potential for generating controversy about socio-economic rather than scientific issues, the use of animal hormones such as bovine and porcine somatotropin to enhance the lean-to-fat ratio of meat seems remote.

1.10 Safety and regulations

In 1974, scientists at the US National Academy of Sciences called for a self-imposed moratorium on certain aspects of experimentation as they became aware of biotechnology's startling potential and its capacity to cause adverse consequences if adequate controls are not applied. Within a year, however, it became clear that further developments in gene technology could not be halted because of the important benefits they offered to society. Adequate regulatory controls designed to ensure the safety of biotechnologically-derived products had to be developed on an *ad hoc* basis due to the virtual absence of quantitative data on the safety of transgenic organisms and their ability to survive in the environment. Since those early days, however, much evidence has accumulated demonstrating the general safety of genetically modified products so that regulations and guidelines for safety assessment can now be developed on a somewhat more rational basis (Engel *et al.*, 1995; OECD, 1993; WHO, 1991). These guidelines and regulations will need continuous revision and updating as new data on GMOs become available.

The UK Advisory Committee on Novel Foods and Processes has recommended that antibiotic marker genes should be eliminated from food products made with genetically modified organisms that have not been inactivated by processing or cooking, as in live yoghurt. Increasingly, alternative and harmless marker genes based, for example, on the ability of an organism to utilize a specific carbohydrate source, are being developed and it is conceivable that these may replace antibiotic marker genes in the future.

In most countries, the release and use of GMOs is strictly controlled and permission has to be obtained from at least one (and very often more than one) government authority (see Chapter 4 for more details). Unfortunately, the ease with which different countries issue permits is variable across the world. In France, permits for field releases are obtainable within one month at minimal cost; in the UK, the process may take up to 6 months and may cost over $3,600 and in Germany, the general political attitudes to biotechnology have been so hostile that permits have been difficult to obtain. Not surprisingly, the difficulties in obtaining permits are reflected in the share these countries have in the total number of releases in Europe: 30%, 17% and 9% for France, UK and Germany, respectively, during the period 1991–1995 (Lloyd-Evans and Barfoot, 1996). However, extensive efforts at harmonization of regulations worldwide are being made now (Chapter 4) and this trend is likely to continue until global consensus on the best way to ensure the safety of genetically modified foods is eventually reached.

It has been suggested that a more restrictive regulatory environment is directly responsible for the disparity in biotechnological developments between Europe and the USA (Lloyd-Evans and Barfoot, 1996). If field releases are accepted as an indicator of biotechnology developments, the European Union lags well behind the USA and Canada: for example, although there were over 500 field releases in the European Union between 1991 and 1995, in Canada there were the same number of releases in 1995 alone (Lloyd-Evans and Barfoot, 1996).

Regulations are perceived by some industrialists as an enemy of a rapidly growing new industry. However, an effective regulatory system serves to reassure the public that they are not being exposed to unassessed and unacceptable risks. Of course, there is room for debate about what form that regulation should take. However, the danger in relaxing the current regime is that it could take only one serious incident to undo all the work done by the regulators in building public confidence in biotechnology.

1.10.1 Labelling

Up until recently, labelling of GM foods has not been mandatory in most countries provided that those foods could be shown to be 'substantially equivalent' to their conventional counterparts (for more details, see Chapter 4). Nevertheless, some retailers and food processors have found that the proactive approach to labelling pays dividends in the long run (Lloyd-Evans, 1994). For example, the Co-operative Retail Society in the UK is selling a vegetarian Cheddar cheese containing chymosin from GMOs with the label: 'Produced using gene technology and so free from animal rennet'.

Some consumer organizations have criticized the concept of voluntary labelling because it is perceived as a way for manufacturers and retailers to reveal only selected information. Other groups, on the other hand, such as the Consumers Association of Canada, have agreed that mandatory labelling of all foods containing GMOs would be unworkable and meaningless. Labelling should be considered on a product-by-product basis. Other ways of informing the consumer such as point-of-purchase stickers and brochures have been suggested and have found favour in the USA (Lloyd-Evans, 1994; Yanchinski, 1996).

Although there is agreement on the consumer's right to know, labelling the products of biotechnology to facilitate informed choice does not achieve its objective if that information is misinterpreted or misunderstood by the consumer.

1.10.2 Patenting of life forms

Another area that is undergoing worldwide review at present is patent legislation in relation to genetically modified life forms. Appropriate property protection for the results of research is vital if the work is to be carried out in the private sector. The USA allows patents on plants and animals (Bizley, 1991). The European Union is moving towards this position with a draft Directive on the protection of biological inventions. However, since there are a number of ethical issues involved, the Directive was rejected by the European Parliament in March 1995. A revised Directive has been proposed but since the ethical issues have remained contentious, the new proposal is likely to have a lengthy journey through the European Commission's corridors before becoming law (Anon., 1996; Bizley, 1991; Crespi, 1991; EFB 1993).

1.11 Consumer acceptance of gene technology in food production

More than 10 years ago in April 1987, the first ever outdoor test of a genetically modified organism took place in a strawberry field in California. A researcher from the company Advanced Genetics Sciences appeared in front of a group of journalists and members of the public wearing a NASA-like protection suit and gas mask, as required by US regulatory agencies at the time. The researcher sprayed a suspension of 'ice-minus' bacteria which had been genetically modified to inhibit the formation of ice crystals on plant leaves onto the plot of strawberries. The following day, the image of the spacesuit-clad scientist was transmitted by all forms of mass media across the globe. Is it really surprising then that much of the public believes that genetic engineering is a dangerous business?

Public misunderstanding of the potential risks associated with gene technology may result in inappropriate rejection of that technology and its applications now and in the future. A consumer may reject the new technology outright as 'a repugnant way of tinkering with nature' (an intrinsic reaction) or may reject specific applications of that technology as in the rejection of plant foodstuffs containing human gene copies (an extrinsic reaction, see also Chapter 3). In either instance, such reactions may result in the loss of potential benefits for both the consumer and the food industry.

Many products of biotechnology which are acceptable in the treatment of terminal disease are not favoured when applied to food. When biotechnology is applied to food, consumers have a choice of whether to accept the technology or not. The exercise of informed choice requires information, and the provision of and reaction to such information will be a key factor in the acceptance or otherwise of food applications of biotechnology.

Various surveys conducted in the USA, Europe and the UK have attempted to assess public opinion with respect to gene technology and its applications (for more detail, see Chapter 2). Most of these surveys have indicated that public understanding of biotechnology is poor, and that genetic modification as applied to food production is perceived as potentially hazardous by the consumer. However, many of these studies were carried out in the 1980s and early 1990s when very few (if any) genetically modified foods were available in the supermarkets. Once GM foods with obvious consumer benefits started appearing on the shelves, the much-feared public opposition failed to materialize and suppliers were generally faced with the opposite problem of insufficient supply in the face of high demand.

It will be essential for the acceptance of food products developed by genetic modification that they offer benefits over conventional products not only for the suppliers (plant breeders, farmers, agrochemical companies, processors or retailers), but also for the consumer.

Some groups or individuals may object on ethical grounds to the consumption of organisms containing copy genes from humans or animals, particularly animals which are the subject of religious dietary restrictions. These genes are referred to as 'ethically sensitive'. For example, Muslims, Sikhs and Hindus have ethical objections to consuming organisms containing copy genes from certain animals such as pigs and cows. Similarly, strict vegetarians would object to incorporation of copy genes from any animal into a plant food (Polkighorne Committee Report, 1993). These and other ethical issues are discussed in more detail in Chapter 3.

1.12 The case for communication

In 1990, the UK government approved the use of a genetically modified baker's yeast without the requirement to label those products made with the new yeast (for more detail on how the yeast was produced, see Chapter 8). An unhelpful and very brief press release issued by the Ministry of Agriculture, Fisheries and Food announcing the approval has since been blamed for provoking negative reactions from consumer and environmental groups. Amidst accusations of government secrecy, the negative consumer reaction eventually culminated in newspaper headlines of 'mutant bread' (Anon., 1990). The producing company, Gist-brocades, has since ceased to market the yeast in the UK and elsewhere (Roller *et al.*, 1994). The case of the GM baker's yeast has provided an example of the possible disastrous consequences of a poor communication strategy whether it be on the part of governments, scientists or producing companies.

While people have a right to question, a right to choose, and a right to manage their own lives, no activist, politician or legislator has produced a public right to understand (IFST, 1996). That has been and remains biotechnology's greatest vulnerability – and greatest challenge.

The exercise of informed choice by the consumer requires accurate and unbiased information and the provision of such information will be a key factor in the acceptance of food applications of biotechnology. What lies behind much of consumer concern with regard to the use of biotechnology in food processing is the lack of a central source of information which has the total trust of the public. Therefore, academic and industrial scientists, professional bodies, learned societies, food retailers, governments and consumer organizations must all play a role in communicating the benefits and pitfalls of food biotechnology to the public. The realization of potential will depend on effective communication not only between scientists but also from the scientist to the lay public, in providing the necessary background information and response to consumer concern (see also Chapter 5 for additional discussion).

In the long run, education of the next generation of consumers is equally important and possibly likely to be more effective than short-term communication efforts with adult consumers. In the UK, a National Centre for Biotechnology Education has been initiated by government and is now funded by donations from charities and income generated through consultancy work for industry. The Centre plays an important role in teacher training and the development of course notes and kits to provide biotechnology education for children of secondary school age. One of their most recent and successful initiatives has been the development of a low-cost and easy to use DNA electrophoresis kit, which can be supplemented with individual modules, the first of which illustrates the action of restriction enzymes used in the isolation and characterization of key genes. By maintaining a low total cost for these kits, their wide application in schools is ensured. A parallel group, Science and Plants for Schools, has very similar aims, and seeks to breed familiarity with approaches used by molecular biologists in plant breeding programmes. Finally, genetics has also been introduced as one of the science elements of the new National Curriculum in British education.

1.13 The future of genetic modification in the food industry

Advances in technology are essential if the food industry is to satisfy the continually rising expectations of the consumer for safer, more nutritious food which is convenient, contains fewer additives and yet is competitive in flavour and price. It is vital for the improvement of the food supply that new technologies, of which gene technology is but one, are taken advantage of as they develop, given safety approval and conformance with any other controls deemed necessary (IFST, 1996). The new biotechnology has the potential to ensure environmentally sustainable supplies of safe, nutritious, affordable and, above all, enjoyable food (Fraley, 1992). However, it has been said by many commentators that the extent to which the potential of gene technology as applied to foods will be realized will depend entirely on public opinion. Another key factor will be trust in the regulatory agencies providing oversight of the technology. In the USA, the FDA and USDA have, in general, the confidence of the public. With the recent issue of BSE in Europe, public trust has been eroded which may be one of the reasons why there is much more concern in Europe than in the USA. Therefore it will be vital to communicate the personal and tangible benefits of biotechnology.

Currently, the benefits of improved agricultural efficiency brought about by biotechnology may not be equally available. However, it is entirely conceivable that plant genetic modifications, now possible without excessive capital investment in expensive equipment, will bring about benefits for the developing world in the future.

References

Anon (1990) *Food Processing*. April.
Anon (1991) ICI Corp. *Abstracts in Biocommerce*, 13 (14), 8.
Anon (1994) Calgene, Campbell Soup and Zeneca. *Biotechnology Bulletin*, March, 8.
Anon (1995) Calgene updates FlavrSavr introduction. *Genetic Engineering News*, June, 26.
Anon (1996a) Legal protection of biotechnological inventions. *Cordis Focus*, 58, 6.
Anon (1996a) Genetically modified tomato puree on sale. *Laboratory News*. Biotechnology News Supplement, March, 1.
Anon (1996a) European project to map the chicken genome. *BBSRC Business* Newsletter, October, 12.
Bauman, D. E. (1992) Bovine somatotropin: Review of an emerging animal technology. Journal of Dairy Science 72, 3432–51.
Bauman, D. E. and Vernon, A. Q. (1993) Effects of exogenous bovine somatotropin on lactation. *Annual Reviews in Nutrition* 13, 437–61.
Bennett, A. B., Chetelat, R. and Klann, E. (1995) Exotic germ plasm or engineered genes. In: Engel *et al.*, 1995. pp. 88–99.
Bleecker, A. B. (1989) Prospects for the use of genetic engineering in the manipulation of ethylene biosynthesis and action in higher plants. In: *Biotechnology and Food Quality*. S. D. Kung, D. D. Bills and R. Quatrano (eds). Butterworths, Boston. pp. 159–65.
Bleecker, A. B., *et al.* (1986) Use of monoclonal antibodies in the purification and characterisation of ACC synthase, an enzyme in ethylene biosynthesis. *Proceedings of the National Academy of Sciences*, 83, 7755–9.
Bridges, I. G., Schuch, W.W . and Grierson, D. (1988) Tomatoes with reduced fruit-ripening enzymes. European Patent Application 0 341 885.
Bizley, R. E. (1991) Animal patents in Europe. *Chemistry and Industry*, 15 July, 505–7.
Chen, T. T. (1994) Making transgenic fish. *Bio/Technology* 12, 249.
Crespi, R. S. (1991) Biotechnology and intellectual property. *TIBTECH* 9, 151–7.
Crawley, M. J. *et al.* (1993) Ecology of transgenic oilseed rape in natural habitats. *Nature*, 363, 620–3.
Do Carmo-Sousa, L. (1969) Distribution of yeast in nature. In: *The Yeasts*. A. H. Rose and J. S. Harrison (eds). Academic Press, London. pp. 79–106.
Dunwell, J. M. 1995. Transgenic cereal crops. *Chemistry & Industry*, September, 730–33.
Dunwell, J. M. and Paul, E. M. (1990) Impact of genetically modified crops in agriculture. *Outlook on Agriculture*, 19 (2), 103–9.
Engel, K-H., Takeoka, G. R. and Teranishi, R. (1995) *Genetically Modified Foods. Safety Issues*. ACS Symposium Series 605. American Chemical Society, Washington DC, USA. 243 pp.
European Federation of Biotechnology (EFB) (1993) Patenting life. *Briefing Paper 1*, June. 4 pp.
Evans, D. A. and Sharp, W. R. (1986) Applications of somaclonal variation. *Bio/Technology*, 4, 528–32.
FDA (1992) Statement of policy: Foods derived from new plant varieties. Food and Drink Administration, *Federal Register*, 57, 22984–3005.
Food and Drink Federation (1995) *Food for our Future – Food and Biotechnology*. Colour brochure available from the FDF, 6 Catherine Street, London WC2B 5JJ.
Food and Drink Federation (1996) *Modern Biotechnology. Towards Greater Understanding*. Colour brochure available from the FDF, 6 Catherine Street, London WC2B 5JJ.
Fox, J. L. (1993a) FDA advisory panel moves Monsanto's BST. *Bio/Technology*, 11, 554–55.
Fox, J. L. (1993b) FDA re-examines biotech policy. *Bio/Technology*, 22, 656.
Fraley, R. (1992) Sustaining the food supply. *Bio/Technology*, 11, 656–59.
Fuchs, R. L. and Astwood, J. D. 1996. Allergenicity assessment of foods derived from genetically modified plants. *Food Technology*, February, 83–8.
Giovannoni, J. J. *et al.* (1989) Expression of a chimeric polygalacturonase gene in trangenic rin (ripening inhibitor) tomato fruit results in polyuronide degradation but not fruit softening. *Plant Cell*, 1, 53.
Graham, A. (1994) A haystack of needles: Applying the polymerase chain reaction. *Chemistry and Industry*, September, 718–21.

Gross, K. C. (1990) Recent developments in tomato fruit softening. Postharvest News Information, 1 (2), 109–12.

Guinee, T. P. and Wilkinson, M. G. (1992) Rennet coagulation and coagulants in cheese manufacture. *Journal of the Society for Dairy Technology*, 45, 94–104.

Hansen, B. and Magnien, E. 1994. *Building Bridges in Biotechnology*. European Commission. 36 pp.

Hassler, S. (1995) Bread and biotechnology. *Bio/Technology*, 13, 1141.

Hobson, G. (1993) The transgenic tomato: The story so far. *Grower*, January, 14.

Hobson, G. and Grierson, D. (1993) Tomato. In: *Biochemistry of Fruit Ripening*. G. Seymour, J. Taylor and G. Tucker (eds). pp. 405–42.

Hooykaas, P. J. J. and Schilperoort, R. A. (1992) *Agrobacterium* and plant genetic engineering. *Plant Molecular Biology*, 19, 15–38.

Houdebine, L. M. (1997) *Transgenic Animals – Generation and Use*. Harwood Academic, 704 pp.

IFST (Institute of Food Science and Technology, UK) (1996a) *Guide to Food Biotechnology*. Published by and available from IFST, 5 Cambridge Court, 210 Shepherd's Bush Road, London W6 7NJ, UK. 70 pp.

IFST (Institute of Food Science and Technology, UK) (1996a) *Genetic Modification and Food. Position Statement*. Published by and available from IFST, 5 Cambridge Court, 210 Shepherd's Bush Road, London W6 7NJ, UK.

IFST (Institute of Food Science and Technology, UK) (1996a) Bovine Somatotropin. IFST Position Statement. Published by and available from IFST, 5 Cambridge Court, 210 Shepherd's Bush Road, London W6 7NJ, UK.

Karp, A. (1993) Are your plants normal? Genetic instability in regenerated and transgenic plants. *Agro-Food Industry Hi-Tech*, May/June, 7–12.

Klein, T. M. *et al.* (1992) Transformation of microbes, plants, and animals by particle bombardment. *Bio/Technology*, 10, 286–91.

Kok, E. J., Reynaerts, A. and Kuiper, H. A. (1993) Novel food products from genetically modified plants: do they need additional food safety regulation? *Trends in Food Science and Technology*, 4, 42–8.

Koziel, M. G., Beland, G. L., Bowman, C. *et al.*, (1993) Field performance of elite transgenic maize plants expressing an insecticidal protein derived from *Bacillus thuringiensis*. *Bio/Technology*, 11, 194–200.

Kramer, M. *et al.* (1992) Postharvest evaluation of transgenic tomatoes with reduced levels of polygalacturonase: processing, firmness and disease resistance. *Postharvest Biology and Technology*, 1, 241–55.

Kramer, M. *et al.* (1990) Field evaluation of tomatoes with reduced polygalacturonase by antisense RNA. In: *Horticultural Biotechnology*. A. B. Bennett and S. D. O'Neil (eds). Wiley-Liss, Inc., New York. pp. 347–55.

Lloyd-Evans, L. P. M. (1994) Biotechnology-derived foods and the battleground of labelling. *Trends in Food Science and Technology* 5, 363–7.

Lloyd-Evans, M. and Barfoot, P. (1996) EU boasts good science base and economic prospects for crop biotechnology. *Genetic Engineering News*, July, 16.

Miflin, B. J. (1993) Bringing plant biotechnology to the market – the next steps. *Agro-Food Industry Hi-Tech*, January/February, 3–5.

Miller, P. D. and Morrison, R. A. (1991) Biotechnological applications in the development of new fruits and vegetables. In: *Biotechnology and Food Ingredients*. I. Goldberg and R. Williams (eds). Van Nostrand Reinhold, New York. pp. 13–29.

Ministry of Agriculture, Fisheries and Food, UK. (1995) *Genetic Modification and Food*. Foodsense series of leaflets. Food Safety Directorate. Leaflet available free of charge from MAFF, London, SE99 7TT.

Morrow, K. J. (1996) Debate on the environmental transfer of engineered plant genes resurfaces. *Genetic Engineering News*, September, 12, 37.

Newton, C.R. and Graham, A. (1994) *PCR – Polymerase Chain Reaction*. Bios Scientific Publishers, Oxford, UK.

OECD (1993) Safety evaluation of foods derived by modern biotechnology. OECD, Paris, France. 80 pp.

Olempska-Beer, Z. S. *et al.* (1993) Plant biotechnology and food safety. *Food Technology*, December, 64–72.

Patel, P.D. (1994a) *Rapid Analysis Techniques in Food Microbiology*. Blackie Academic & Professional, Glasgow.
Patel, P. D. (1994b) The use of DNA fingerprinting in food analysis. *Food Technology International Europe*, 171–5.
Pfeiffer, N. (1994) FDA OKs Calgene's Flavr Savr tomato for marketing in supermarkets in the U.S. *Genetic Engineering News*, June, 1, 31.
Redenbaugh, K. *et al.* (1992) *Safety Assessment of Genetically Engineered Fruit and Vegetables: A Case Study of the FlavrSavr™ Tomato*. CRC Press, Boca Raton, Florida, USA. pp. 267.
Redenbaugh, K. *et al.* (1994) Regulatory assessment of the FlavrSavr™ tomato. *Trends in Food Science and Technology*, 5, 105–10.
Rissler, J. and Mellon, M. (1994) No commercial gene-altered crop approvals until Fed Government assesses the ecological risks. *Genetic Engineering News*, February, 4, 12.
Roller, S., Praaning-van Dalen, D. and Andreoli, P. (1994) The environmental implications of genetic engineering in the food industry. In: *Environmental Issues and the Food Industry*. J. Dalzell (ed.). Blackie Scientific, London. pp. 48–75.
Schuch, W. (1994) Improving tomato quality through biotechnology. *Food Technology*, November, 78–83.
Sheehy, R. E., Kramer, M. and Hiatt, W. R. (1988) Reduction of polygalacturonase activity in tomato fruit by antisense RNA. *Proceedings of the National Academy of Sciences*, 85, 8805.
Shewry, P.R. *et al.* (1995) Biotechnology of breadmaking: Unraveling and manipulating the multi-protein gluten complex. *Bio/Technology*, 13, 1185–90.
Smith, C. J. S. *et al.* (1988) Antisense RNA inhibition of polygalacturonase gene expression in transgenic tomatoes. *Nature*, 334, 724.
Smith, C. J. S. *et al.* (1990) Inheritance and effect on ripening of antisense polygalacturonase genes in transgenic tomatoes. *Plant Molecular Biology*, 14, 369–79.
Swaminathan, B. and Feng, P. (1994) Rapid detection of foodborne pathogenic bacteria. *Annual Review of Microbiology*, 48, 406–26.
Taylor, P. and Thomas, J. (1994) Genetically modified plants – benefits, concerns and control. *Lipid Technology*, July/August, 91–2.
van Wagner, L. (1993) Jurassic Park meets the food industry. *Food Processing*, December, 79–80.
Wasserman, B. P. (1990) Expectations and role of biotechnology in improving fruit and vegetable quality. *Food Technology*, 44 (2), 68.
WHO (1991) Strategies for assessing the safety of foods produced by biotechnology. Report of a Joint FAO/WHO Consultation. WHO, Geneva. 59 pp.
Yanchinski, S. (1996) Canadian health agency OK's Monsanto's genetically engineered potatoes. *Genetic Engineering News*, February, 24.
Zambryski, P. C. (1992) Chronicles from the *Agrobacterium*-plant cell DNA transfer story. *Annual Reviews of Plant Physiology and Plant Molecular Biology*, 43, 465–90.

2 Consumer perceptions of modern food biotechnology

LYNN J. FREWER and RICHARD SHEPHERD

2.1 Introduction

The significant economic and social benefits of modern biotechnology may not be realized if consumer acceptance issues are not adequately addressed (Stenholm and Waggoner, 1992). Public reaction is a crucial factor in developing and introducing biotechnology (Cantley, 1987; De Flines, 1987). The issue of consumer confidence in novel products, whether from the point of view of safety or perceptions of quality, must be answered within the wider social context in which the technology is embedded. The exploitation of biotechnology to its full extent is likely to depend on public acceptance of a range of issues including perceptions of ethical and socio-economic impacts, as well as food safety. Cross-cultural differences in acceptance are likely to exist, as well as individual differences within specific populations.

Most experts recognize that public knowledge and perceptions of biotechnology must be systematically evaluated (Hoban and Kendall, 1993). Commercially viable products may not necessarily be marketed if there is evidence that public rejection will occur. In addition to the problem of wasted R&D resources, it is possible that a negative public reaction to an individual product of the technology will spread to other, hitherto acceptable products. Indeed, public perceptions of the technology overall may be focused on a single example of inappropriate application. It is claimed that biotechnology has replaced nuclear power as the symbol of 'technology out of control', with little reference being made to the many positive benefits which are the result of its continued development and application (Nelkin, 1995). Great care should therefore be taken in systematically and effectively addressing consumer issues.

It has been noted that emerging technologies reach the attention of planners, forecasters and policy makers in a time series (Bauer, 1995). Comparing nuclear power, information technology and biotechnology, the potential impact of these technologies were assessed for the first time in OECD policy reports in 1956, 1971 and 1982, respectively. While nuclear power has been the focus of public reaction, public opinion on biotechnology is only just beginning to crystallize – the direction that this opinion takes is likely to determine the future direction of policy surrounding the technology as well as the path of future developments. Jasanoff (1995) has commented that, over the last twenty years, 'biotechnology has moved from a research programme that aroused misgivings even among its most ardent advocates to a flourishing industry throughout Europe and the United States'.

Jasanoff (1995) has compared the differences in the regulations adopted in the UK, the United States and in Germany to control biotechnology. In the UK, regulation has been predominately 'process driven', and has taken some account of the social and political framework surrounding the technology. In the United States, the regulatory framework has developed on a case-by-case basis, and is strongly driven by technical risk estimates linked with individual products. By contrast, Germany's regulatory framework has taken more account of the political, social and ethical pressures linked to the technology. However, it is notable that all three regulatory approaches appear to have converged in terms of what is acceptable. For a more detailed review of regulations concerning biotechnology, the reader is referred to Chapter 4.

However, regulation (whether internationally convergent or divergent) has not solved the consumer acceptance issue, despite the high expressed public need for such control (Frewer and Shepherd, 1995). Various pressure groups specifically opposed to the technology are beginning to form in parallel with the development of policy at a regulatory level, particularly in Europe. The development of such pressure groups reflects the need for a policy which addresses factors other than technical risks and benefits, but which also takes into account the many ethical and socio-economic factors which are of importance to the general population. Scientists have tended to focus on the risks associated with specific biological and medical hazards, which is directly reflected in national regulation. The public, on the other hand, have been more concerned about social, ethical and ecological factors (Frewer *et al.*, 1997a; Knudsen and Ovesen, 1994). For consumer acceptance of the technology and its products to occur, all aspects of public concern must be addressed.

2.2 Knowledge and public understanding

Research in the United States (Hoban and Kendall, 1992) and Europe (Marlier, 1992) has indicated low public awareness of biotechnology. This does not prevent the public from making risk assessments which assume that genetic modification is risky, particularly when applied to food production (Sparks, Shepherd and Frewer, 1994). However, increased public understanding does not necessarily facilitate acceptance of the technology (Hamstra, 1991). Better public understanding facilitates *informed choice* in the selection of genetically modified products, and is therefore an important facet of the development of the technology. In addition, labelling the products of the technology may provide a high-risk signal unless the public understand what the labelling means.

Michael (1992) comments that the public differentiate between science as an abstract entity (science in general) and as an activity directed at specific events or problems (science in particular). Consumer selection of products of new technologies is in the latter category (Bauer, 1995). Acceptance or rejection may, in part, be the result of public misunderstanding of the relevant scientific questions. In addition, acceptance/rejection of new technology may have 'spatial' and 'cluster' effects. Resistance in one location is not necessarily indicative of resistance in another, resulting in differential national exploitation of the technology (Hagerstand, 1967). This may confer a technological disadvantage to some countries in terms of commercial exploitation and economic development.

While there has been an historical view that the public should be 'educated' to accept the technology, there is some evidence that increased knowledge and understanding may serve to polarize attitudes, an effect which may be very dependent on the prior attitudes held by the population (Frewer, Howard and Shepherd, 1997). Foreman (1990) has noted that communication about biotechnology may result in conflict if the risks and benefits from the technology do not accrue equally between different groups within the population. For example, if the public believe that the benefits are only applicable to industry, but the risks will impact on the environment and affect the whole population, then low public acceptance of the technology might result. This type of effect is likely to occur even if there is low public understanding regarding the long-term and equitable benefit of technological development. To understand how the public is likely to react to new developments of genetic modification, it is necessary to try to understand the way people psychologically characterize risks and benefits resulting from its application.

2.3 The importance of social context

Risk perception is socially constructed, and not defined by what has traditionally been considered 'technical' risk estimation. Public demands for a complete absence of risk in society have confused technical experts, who have tended to attribute such an effect to public 'ignorance'. However, 'scientific' objective reality may be very different from that found within a social context. Lay beliefs about the reality of risk associated with any technology are as likely to depend on perceptions of social conflict and credibility, as the actual levels of danger facing an individual member of society from any given hazard. Social constructionists focus on how power works in framing the terms of the debate about risks, and the influence that cultural factors have in the processes associated with risk perceptions. In the case of technological hazards, the risk debate is embedded in the social and political meanings of technologies. Alternative arguments to this technological approach focus on the possibilities of low risk, low technological alternatives. Risk perceptions are constructed at a social level beyond technical arguments, and, at least for some members of society, focus on alternatives to the 'hard technology' approach.

To effectively communicate risk information, it is crucial that the social construction of risk perception is taken into account. Jasanoff (1993) has noted that, today, most risk analysts are in agreement that facts and values tend to merge together, particularly when hazards are high in uncertainty. The importance of social context may increase if there is a potentially catastrophic consequence resulting from the hazard should it occur, even if it has a low probability of occurrence.

The social construction of risk perception can be illustrated by examination of the disparities between expert and lay concerns regarding the nature and relative importance of different hazards. For example, in Sweden, it has been found that risks from nuclear radiation were judged to be wholly unacceptable by the public, although the same risks were dismissed by experts, who regarded them as negligible (Sjoberg and Drottz-Sjoberg, 1994). In the USA, the lay public and experts differ not only in their opinions of risk magnitudes associated with the handling of nuclear waste, but also in the conceptualization of what types of risks represent a serious threat (Flynn, Slovic and Mertz, 1993). In the case of the 'new genetics', experts have been found to be much more positive than lay members of the public towards the application of new techniques in genetic screening (Michie *et al.*, 1995).

An additional complication is that risk information provided by 'experts' is equally likely to be influenced by the social context in which the communicator finds him or herself. Scientists in universities or local governments typically perceive the risks of nuclear energy and wastes as greater than scientists who work as business consultants, national government or for private research establishments (Barke and Jenkins-Smith, 1993). It would not be unreasonable to predict that a similar effect might occur within the area of the new biotechnology.

The reason for such disparity between public risk perception and technical estimates of risks is that the public tend to characterize hazards using psychological constructs which are very different from those which simply refer to probabilities of occurrence. The factor analytic model developed by Slovic and his colleagues describes various hazards in terms of two dimensions – the first factor is associated with new, involuntary, uncontrollable risks, the second with fatality and catastrophe (Fischhoff *et al.*, 1978; Slovic, Kraus and Covello, 1990). 'Number of people exposed' was later added to the model as further research indicated that it was likely to be important

(Slovic, Fischhoff and Liechtenstein, 1980). It is likely that this indicates that 'societal threat' is important as a determinant of people's reactions to risks. A similar structure of risk perception has recently been reported specifically for food-related hazards (Sparks and Shepherd, 1994). Using this descriptive framework, genetic engineering in food production has been described as an extremely unknown hazard of moderate severity, relative to other food-hazards, and indeed hazards in general. Public acceptance is more likely to relate to this very different conceptual framework than the more common cost-benefit analysis proposed by experts.

Food risks which are associated with processing technologies seem to be particularly perceived as relatively high in risk compared to the benefits which people associate with their application, as shown in Figure 2.1. Here it can be seen that the ratio between the relative perceived risk and benefits for a range of technologies is greatest for the food-related technologies, which tend to group together close to the most negatively viewed technology, that of nuclear energy. By contrast, solar energy is perceived as high in benefit relative to risk (Figure 2.1).

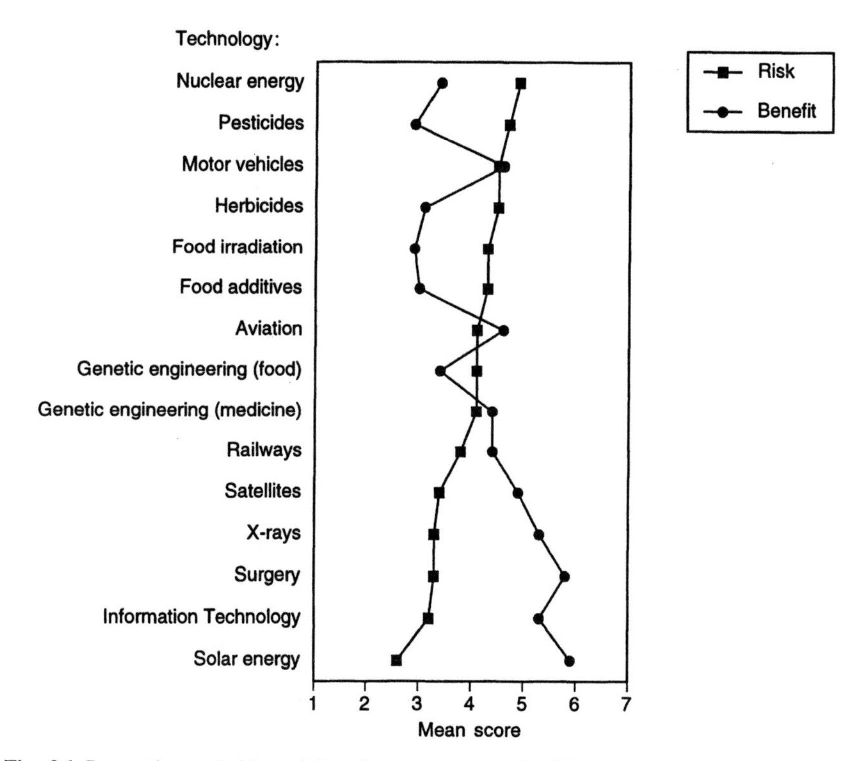

Fig. 2.1 Perceptions of risk and benefit associated with different technological hazards. Adapted from Frewer, Howard and Shepherd (1995), by permission of MCB University Press.

2.4 The international picture

It has been estimated that, from 1984 to 1994, more than 70 000 people have been asked what they think about biotechnology (Zechendorf, 1994). Research has been conducted primarily in the USA, Europe and Japan, with very little information being available for populations outside of the developed world. The lack of information about attitudes in developing countries is clearly an area which urgently needs to be addressed, partly from the perspective of 'democratizing' the future development of the technology on a world-wide basis, and partly because many of the socio-economic and ethical issues are focused around applications which will be applied within developing countries, or at least have economic impacts in these areas.

Comparison of surveys constructed in different countries at different times using different methodologies creates difficulties in making general statements regarding overall trends in attitudes. Even the European 'Eurobarometer' survey used different questions in the 1991 and 1993 waves of cross-national surveys (Eurobarometer, 1993). Furthermore, there may be cross-cultural differences in the absolute levels of response. For example, there may be a tendency in a particular country to rate all questions relating to risk perception (and not just those associated with biotechnology) as higher than in another country being used in the comparison, depending on other recent political and technological events within the different communities (Zechendorf, 1994).

Nevertheless, some key similarities and dissimilarities between different cultures emerge from the data, although interpretation must be made with care. In the United States, public understanding of the technology is low, but so is opposition to the application of the technology. In Northern Europe, the public tend to be better informed about the technology. Ethical and moral concerns are of particular importance in the UK (Frewer and Shepherd, 1995), and in Northern Europe (Zechendorf, 1994). Perceptions of 'tampering with nature' and 'unnaturalness' appear to be a focus of opposition in the US (Hoban and Kendall, 1992) and the UK (Frewer, Howard and Shepherd, 1997). In Southern Europe, (and Ireland) risk perceptions are lower and acceptance of food-related applications higher. Southern Europeans are more positive about genetic manipulations of animals compared to plants and microorganisms (Zechendorf, 1994), which is not the case in Northern European countries such as The Netherlands (Hamstra, 1991) and the UK (Frewer, Howard and Shepherd, 1997). The Japanese perceive high levels of risk in food biotechnology, despite being more informed about the science underpinning its application, but have more positive attitudes towards pharmaceutical developments (Hoban, 1996b).

There are problems in extrapolating from these findings to make predictions about world-wide acceptance. Most importantly, attitudes are likely to become more focused as products become available to the consumer, and the tangible benefits of the technology become obvious (Frewer, Howard and Shepherd, 1996). Secondly, consumer reactions are likely to be dependent on the desirability of benefits associated with very particular products. Thirdly, individual differences in attitudes are likely to exist even within very specific populations (Frewer *et al.*, 1997a). It seems most appropriate to consider specific applications on a case-by-case basis, and for individual companies to engage in systematic market-testing of products in the development stage in all countries where there is likely to be a potential market for products.

Switzerland provides a good example of the use of referenda in the identification and public assessment of new technologies and their future development. Buchmann (1995) has noted that the Swiss population has mixed public opinions about the new biotechnology, positive in terms of attitudes to progress, but negative in terms of the potential for misuse, particularly within the context of self regulation by scientists and industry. A referendum regarding increased constitutional regulation of biotechnology was held in 1992, which resulted in a 'yes' vote for increased regulation, which Buchmann interpreted as indicative of a strong signal to interested institutions to consider citizens' concerns in the future development of the technology. It is suggested here that such increased public participation is likely to increase consumer acceptance of its products, as the wishes of the public are incorporated into the legislative framework, and made public prior to product development.

Research into consumer attitudes has been particularly prolific in The Netherlands, perhaps due to the fact that consumer concern is high in this country, and that the economy is largely dependent on the export of agricultural produce. Research has tended to focus on specific products which are under development, rather than assessing very general attitudes to the technology. Hamstra (1992) reported that consumers who reject one genetically modified food product hold a generally negative attitude towards the technology, and are unlikely to accept other products. In contrast, research from the UK has indicated that attitudes are likely to become more focused by exposure to specific products (Frewer, Howard and Shepherd, 1996), and are more likely to be positive towards applications using plants and microorganisms, as opposed to human genetic material or animals. Hamstra (1992) has, however, added the *caveat* that products with a clear advantage for consumers may counterbalance factors leading to rejection. A clear benefit (for example, reduced use of pesticides), is likely to facilitate acceptance. Heijs Midden and Drabbe (1993) report that, in The Netherlands, attitudes appear to be focused by the 'area of application' and the 'purpose' or 'end product' of the application, with a generally neutral attitude towards applications in the food industry (although the low level of public knowledge at the time of data collection may have influenced the results).

As products come 'on stream', it is highly likely that the first products to appear will have the greatest impact on consumer attitudes. Given that some applications are less acceptable than others (for example, genetic manipulation involving animals or human genetic material), the order in which products become available to the public is critical, as the earlier ones will have the greatest impact on public perceptions, and are likely to influence the formation of public attitudes.

2.5 Risk perception and consumer reaction

How is genetically modified food characterized in risk perception terms relative to other food-related hazards? In a recent study, respondents were asked to rate different food-related potential hazards on pertinent issues such as the seriousness of the hazard, control, voluntariness and whether the hazard is likely to be fatal. The mean responses were analysed by principal components analysis. The first component accounted for 45% of the variance and was labelled as 'severity', while the second component accounted for 33% of the variance and was labelled 'known risk'. Genetically engineered food was perceived as being a moderately severe risk, but one which was very unknown, relative to the other food hazards included in the study (Sparks and Shepherd, 1994).

However, consumer acceptance of the food products of genetic modification are unlikely to be determined by attitudes to the technology overall. Rather, acceptance will be determined by recognition of the tangible benefits of specific applications of the technology (Frewer *et al.*, 1997b). Previous research has indicated that public attitudes towards genetic engineering are likely to change dramatically when the application is tied to a specific goal (Powell and Griffiths, 1993). Even those applications seen as risky, or otherwise perceived to be negative, may be accepted if the application is also seen to be necessary or useful. This effect is likely to be cross-culturally applicable even if the focus of objection differs between the different cultures. In general, research has indicated that where experimental research is tied to a particular product or product type, with distinct benefits, consumer reactions are likely to be very closely linked to the characteristics of the product itself (e.g. Hamstra, 1991; Frewer, Howard and Shepherd, 1997).

Finally, the importance of ethics cannot be ignored. Ethics is defined as 'a set of standards by which a particular group decides to regulate its behaviour – to distinguish what is legitimate or acceptable in pursuit of their aims from what is not' (Godown, 1987). Some ethical arguments (equivalent to intrinsic arguments in moral reasoning) may focus on the development of the technology *per se*, and relate to ethical opposition to the technology because it is fundamentally wrong. Extrinsic arguments against the technology are related to the consequences of application, and as such are more closely related to perceived risks or outcomes of applications. From the pattern of public attitudes which have emerged to date, it seems likely that public ethical concerns are very much focused on specific applications, and that legislative control related to ethics should be examined on a case-by-case basis (Straughan, 1992). It is of course possible that ethical standards will change with time (consider, for example, the changing ethical code on issues like abortion, or animal welfare), and ethical decisions may need to be revised if contemporary standards are judged not to be adequate. For further discussion of the moral and ethical issues in food biotechnology, the reader is referred to Chapter 3.

2.6 Credibility of the risk regulators

Trust in risk information about genetic modification may be as important a determinant of consumer reactions as the content of the risk information. The 1993 Eurobarometer poll which investigated European attitudes towards biotechnology and genetic engineering indicated that industry and government were distrusted as accurate information sources (Eurobarometer, 1993). Medical applications of genetic engineering are regarded as more acceptable than food-related applications by the public, in terms of perceptions of risk and benefit, and ethical concerns relating to the technology (Frewer and Shepherd, 1994). This is possibly partly due to the greater level of trust conferred onto the managers of the medical technologies as opposed to those associated with industry (Slovic, 1993). The information which is communicated by the different institutions (medical organizations versus industry) is also likely to be differentially trusted (Martin and Tait, 1993). In the case of gene technology, there is evidence that there is little trust in either the biotechnology industry or in the governmental organizations which regulate it. Institutional qualities such as accountability and trustworthiness are extremely important in determining public reactions to an event or development (Wynne, 1989). Indeed, the level of public trust and confidence in institutions and individuals involved in risk management has been shown to be a major factor in the evolution of disputes about risk decisions (Kasperson, Golding and Tuller, 1992).

2.7 The role of the media in consumer acceptance

Marlier (1992) has reported that the key source of information used by Europeans to gain information about genetic modification are the media. In the UK, the quality media are the most trusted sources of information, although the tabloid press is highly distrusted (Frewer and Shepherd, 1994).

Similarly, research from the United States and Japan has indicated that the media are highly influential in the crystallization of consumer knowledge and attitudes about food produced by biotechnology (Hoban and Kendall, 1993; Hoban, 1996a,b). Many information campaigns assume that perceptions and attitudes will align with scientific judgments through the mediating influence of media exposure (Liu and Smith, 1990).

There are two major issues associated with media coverage of biotechnology:

1. What is the extent of coverage?
2. What is the content of coverage?

Mazur and Lee (1993) have argued that negative public reactions are not the result of criticism or negative bias in news coverage, but rather it is the amount of coverage which is the primary driver of negative public reactions. The theory of 'agenda setting' assumes that the media do not directly influence what the public thinks, but are successful in making issues salient to the public (McCombs and Shaw, 1972). In support of this hypothesis, Mazur (1987) has noted that public concern about environmental hazards rises and falls with the volume of reporting, even if the overall tone of the coverage is generally positive. From a cognitive psychology perspective, simple but frequently repeated images are likely to result in the formation of an 'availability heuristic', or establishment of an internal rule that signifies potential danger (Tversky and Kahnemann, 1981). However, content must also be examined, in order to understand how the available information is likely to impact on public attitudes. Furthermore, it is unlikely that an article about gene technology published on the science page will have the same impact on the public as front page headlines presenting the technology in a 'crisis' context.

Examination of the type of risk reporting associated with different food-related hazards in the UK has indicated that different food hazards are linked with different types of risk information (Frewer, Raats and Shepherd, 1993/1994). Figure 2.2 illustrates that food additives, for example, are frequently presented in the media, but are associated with very little concrete risk information – rather they are presented as a hazard to be avoided, with little additional qualifying risk information associated with the hazard. Microbiological risks are associated with much more risk information, which tends to be more statistical, rather than qualitative in nature. Biotechnology tends to be presented as an unknown risk, and is associated with conflicting statements originating from different actors in the risk debate, and qualitative and descriptive risk statements.

Extensive study of media reporting of genetic engineering has been conducted in Germany. Ruhrmann (1992) has examined the question of media reporting of genetic modification from the perspective of examples of the technology reported, the different risk perception factors reported, and how

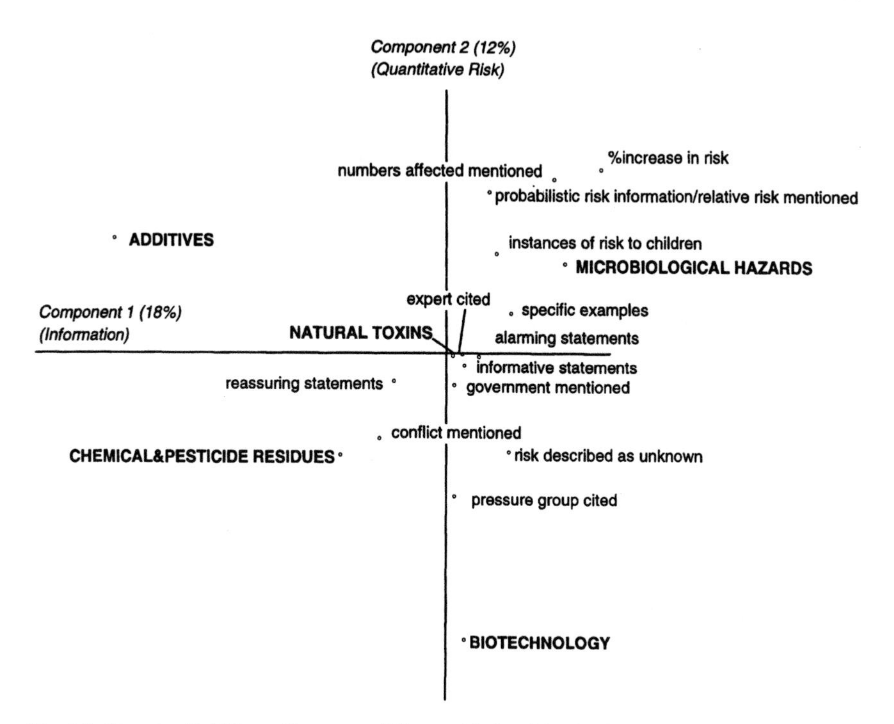

Fig. 2.2 How the British media report different kinds of food-related hazards. From Frewer, Raats and Shepherd (1993/1994), by permission of Oxford University Press.

reporting relates to the political development of the technology. The analysis covered the period from 1988 to the beginning of 1990. Medical research and development work was the most frequently reported story 'type', although economic competition, genome mapping, plant and animal breeding, ethical debate, policy and regulation and eugenic potential were also important. It was found that greatest emphasis was placed on the catastrophic potential of the technology. Less attention was paid to the idea that the associated risks were potentially unknown. In articles dealing with the broader consequences of gene technology a hypothetical risk was assumed, whereby the risks were presented as undefinable or unmeasurable, but which were likely to occur in the future if safety precautions were not taken at the present time. In political terms, the explicit opponents of gene technology were presented almost four times as often as the technology's proponents, emphasizing that conflict is an important factor in news worthiness of the technology.

Kepplinger (1995) has argued that one of the basic questions regarding media coverage of genetic engineering is whether the views held by scientists are shaped by science writers, and whether their views shape the coverage of news reporting. Scientists were found to have more positive attitudes compared to journalists, who tended to have little scientific knowledge regarding the technology. Most articles appeared in political sections of the newspapers, rather than scientific sections. In the political sections, more space was devoted to those actively opposing the technology. Quotes from the various actors in the risk debate tended to be selected to support the views of the journalists themselves.

More research is needed to understand how media information influences both public attitudes about biotechnology as a developing science, and their attitudes as consumers of its products.

2.8 The development of effective communication about biotechnology

All those with an interest in developing and applying biotechnology are likely to benefit from the development of effective communication models. Effective communication will ensure that consumers are making decisions based on accurate information about processes and products.

For many consumers, food labels may be one of the most important sources of information about gene technology. As the requirements for labelling are likely to become more stringent, it becomes imperative that consumers understand what the information contained on the label means, rather than dismissing the product because it is seen as the result of a new and unnatural technology. Furthermore, control by individuals may be linked to labelling of products and without such labelling people may feel they have little opportunity for personal control. However, labelling will have little impact without public understanding of what the labels mean. Therefore, there is a clear need for effective communication strategies to facilitate public understanding of gene technology.

Proactive (rather than reactive) information provision should be made, particularly regarding ethical concerns, potential risks and control of biotechnology. The establishment of a dialogue between the public and experts will enable key public concerns to be addressed, as well as those directly salient to risk issues. Some of the communication strategies adopted to date are reviewed in Chapter 5.

Smink and Hamstra (1996) have made an extensive study of the labelling requirements of Dutch consumers regarding the products of genetic engineering. While it is reported that the labelling question in itself is not the main issue for most consumers, a substantial majority would like to be informed about product-related risks and benefits. Labelling by itself is unlikely to provide all the information required by the consumer. For example, alternative and expanded information sources, such as consumer helplines, might reassure those members of the public who require additional information. There is a danger that differential labelling policies might develop whereby some products are labelled, and others are not.

Unless there is consensus between industry and risk regulators as to what is to be labelled, and the consumer is aware of this agreement, public distrust in manufacturers and risk regulators might result. Conflict may signal that there is a hidden risk of which the public is not aware.

Urban and Hoban (personal communication) have suggested that the most efficient way of altering negative risk perceptions of biotechnology is not to increase the publicly available amount of information on biotechnology, but to try to influence a more general pessimism towards technology. However, it would be a mistake to attempt to promote an overall emotional and optimistic perception of biotechnology, as this might be perceived as being unduly persuasive and aimed at promoting the vested interests of various stakeholders.

2.8.1 What should be communicated?

Effective communication should be driven by what the consumer wants to know about a specific application of the technology, as well as the scientific facts surrounding the technology. For example, consider the case of the consumer concerned about potential environmental effects of genetically modified rapeseed. In this instance it is of little relevance to discuss the scientific process by which the plant is genetically modified if additional information about potential environmental impact of the crop is not provided. A comparative example can be drawn from the health information literature. In the case of AIDS transmission, the concern of medical authorities focused on the low percentage of the population who knew that transmission of the disease was caused by a virus. However, it is not public understanding of scientific medical information, but rather effectively communicating information regarding preventive behaviour which is important (Fischhoff, Bostrum and Quadrel, 1993).

Effective communication is likely to enhance public trust in industry, risk regulators and future technological developments. Increasing perceptions of industrial proactivity in public communication at a level which is meaningful to the consumer is likely to facilitate acceptance of novel products. Risk communication is a relatively new but rapidly expanding area, and at present has its basis in a variety of different disciplines. A review of key developments in the area is provided in Fischhoff (1995). The rapid growth of the risk communication literature reflects the frequent social conflicts surrounding the management of environmental hazards (Plough and Krimsky, 1987). Clearly, there are major insights which can be extrapolated from this research to the development of effective communication strategies about genetic engineering.

Covello and his co-authors (1986), reviewing the literature in the area, identified four types of objectives of risk communication:

1. behaviour change and protective action
2. disaster warnings and emergency information
3. information and education
4. joint problem-solving and conflict resolution

The last two of these are directly relevant to the development of risk–benefit communication associated with gene technology. The 'public dialogue' approach is most likely to be effective in risk–benefit communication, where the emphasis is not just on safety, but also on improvements in quality of life and economic advantages. In the case of genetic modification, it is important to take account of the reasons why the technology is being developed (the benefits) as well as issues which are likely to cause concern.

It has been pointed out that one of the problems associated with effective risk communication is the potential for divergent responses within culturally divergent societies (Vaughan, 1995). For example, Tyler and Griffin (1991) distinguish between two types of 'justice' orientations in the equity of outcomes associated with different technological hazards. The first of these, 'distributive' justice, refers to equity in the distribution of risks and benefits across the whole population. That is, are particular sectors of the population disproportionately at risk, while others receive a disproportionate share of the benefits? The authors point out that in some instances, disproportionate risk exposure may be associated with disproportionate receipt of benefits, implying that these factors must be taken into account in the development of risk–benefit communication.

If the benefits of biotechnology are perceived to apply only to industry and commercial interests, whereas risks are believed to be borne by the consumer, then communication is likely to be ineffective. Procedural justice refers to the fairness of processes by which decisions are reached, which directly implies the need for a participatory communication process, rather than the 'top-down' approach which has been used in the past. Vaughan (1995) has emphasized the fact that many communities are concerned about how risk management decisions are arrived at – about questions of procedural justice. Particularly in the case of environmentalist issues, non-governmental organizations have an expectation of involvement in the discussion of meaningful alternatives in the deliberation process (Capek, 1993), and involvement of such groups might be best sought early in the development process.

2.8.2 Source credibility

It has been shown in many European countries, as well as in the USA and Japan, that industry and, to a more variable extent, governments, often lack public trust and credibility, both as risk regulators and communicators

(Martin and Tait, 1993; Hoban, 1996a,b). Experimental work has been conducted to attempt to understand the underlying psychological constructs which lead to differentiation in source credibility (Frewer *et al.*, 1996). Expertise in itself does not lead to trust; rather, trusted sources are seen to be characterized by multiple positive attributes, such as having public welfare as a primary organizational concern, being factual and knowledgeable (although not necessarily expert) and having a 'good track record' of information provision. The research has also shown that complete freedom does not lead to trust, but is linked with 'amplification' effects whereby risk information is seen to be exaggerated. Sources which possess moderate accountability are seen to be the most trusted. To maximize trust, a source needs to be perceived as being moderately accountable. Too much accountability, and the information is seen to be distorted to conform to vested interests. Too little accountability, and the information is seen to be exaggerated.

An additional factor which must be taken into account when assessing questions linked to credibility is the social context in which the information is presented. This relates to the nature of the hazard, and the context in which information is presented. For example, self-reported trust in sources of information about genetic engineering in a hypothetical situation (where no information is presented) may not equate with behavioural responses to actual information attributed to a particular source (Frewer and Shepherd, 1994). In the case of genetic engineering in food production, it has been found that actually presenting information reduced trust for highly credible sources so that it equated with that reported for the less trusted source.

It has been noted that communication processes perceived as exclusive (where there is no attempt at interactive communication with the public) can result in distrust in the communicator, as well as opposition to risk management decisions (Capek, 1993).

2.8.3 *Prior attitudes and source characteristics*

Both theoretical social psychology (Eagly and Chaiken, 1993) and applied risk perception research (Vaughan, 1995)have indicated that prior attitudes to potential hazards are likely to be an important determinant of reactions to incoming information. If the information is presented very positively towards a technology, negative attitudes may become more negative and positive attitudes more positive.

There are problems in making predictions about the long-term impacts of effects such as these. Firstly, at present, public attitudes are clearly very uncrystallized – if there is acceptable evidence that the technology is beneficial, a greater percentage of the population are likely to formulate positive attitudes. However, should a serious error be made in technological application and development, this is likely to act as a signal to the public that the new biotechnology is dangerous and uncontrollable, and focus attitudes in a negative direction. This would influence public responses to information in a very different way compared with a situation where a similar error had not been made.

2.8.4 *Individual differences*

Not all members of the public are likely to react in the same way to genetic modification of foods. In the UK, women are more likely than men to object to animal-related applications, or applications involving human DNA if the use of the technology is presented in very general terms. Similarly, in the USA and Japan, women have generally less positive attitudes towards and less awareness of biotechnology than do men (Hoban and Kendall, 1993; Hoban, 1994, 1996a,b). However, these gender differences tend to disappear if specific applications with more obvious tangible benefits are presented to respondents. Environmental concern is a good predictor of negative reactions of individuals to animal-related applications, unless the reasons for the modification are explained (Frewer *et al.*, 1997a).

In general, it is more useful to try to understand differences in attitudes within the wider socio-economic structure in which individuals find themselves. Relative poverty is likely to be important. For example, individuals having fewer socio-economic resources have been shown to have lower judgements of control over many risk outcomes (Mirowsky and Ross, 1986). In the area of occupational health hazards, poorer socio-economic groups tend to adopt a more 'fatalistic' orientation towards future negative health outcomes. It is not at present known whether socio-economic status is consistently influential in determining the effectiveness of risk communication about biotechnology. Intuitively, one might expect that benefits such as lower priced food products might be more attractive to socio-economically disadvantaged groups.

2.8.5 *The consensus conference model*

One method which has been proposed as an effective means of incorporating lay perspectives into the development of policy associated with new sciences and technologies is that of the consensus conference (Joss and Durant, 1995). A consensus conference is a meeting in which a group of lay people invite technical experts, as well as proponents and opponents of a controversial technology, to answer specific questions; the lay panel then assesses the experts' responses, reaches a consensus about the issue under discussion, and reports the conclusions at a press conference. The goal of such an exercise is to create a two-way dialogue between the public and technical experts. The approach has been very influential in Denmark (where it was developed) and The Netherlands, in terms of both stimulating public debate and influencing policy. A similar approach, the 'citizens' jury' has been used in Germany and the United States (Stewart, Kendall and Coote, 1994).

The first UK consensus conference was held in 1994 on plant biotechnology and was successful in that agreement regarding the wider social issues was reached by the lay panel. Although the technology was generally regarded as being acceptable in this instance, the ethical, social and political issues were highlighted as important (National Consensus Conference, 1994). Limitations of the approach include those linked to developing the media exposure to reach a national audience (one competing high interest news event will ensure that media coverage is minimal, despite careful planning). Furthermore, the small number of lay people involved in a consensus conference (who may not necessarily have views which align with the population as a whole), mean that the final report cannot be nationally representative. Finally, the need to develop consensus within the group does not take account of the fact that individual differences are likely to be highly influential in determining individual reactions to emerging technologies.

2.9 Future research

The major focus of future research is likely to be the development of effective communication strategies about the new biotechnology, and issues surrounding the debate about public understanding of science. Understanding of causation of individual differences (both inter- and intraculturally) is likely to be fruitful both from the perspective of predicting probable consumer acceptance of specific products, and developing the academic basis of attitude research.

One area is emerging from the literature – that it is imperative to assess the long-term impacts of effective communication strategies on attitudes. Systematic study of the temporal persistence of attitudes has been rare (Eagly and Chaiken, 1993), and there are different models to predict the long-term effects of information provision on attitudes. The 'sleeper effect' refers to the situation where attitude change is delayed, because an unfavourably evaluated external cue (such as distrust in the source) becomes dissociated from it over time (Hovland, Janis and Kelly, 1953). Against this, the concept of associative interference would predict that while attitude change reduces over time, this effect is likely to be highly dependent on the contextual cues surrounding the initial message (Lieppe, Greenwald and Baumgardner, 1982). The long-term impact of source credibility on attitude formation needs to be investigated within the context of both general attitudes to the new biotechnology, and for particular examples of the technology which are coming 'on-stream'.

2.10 Conclusions

Much research has been conducted into the issue of consumer acceptance of biotechnology. However, more focused research into the area is required as attitudes change, particularly as new products come onto the market, something which is likely to focus public attitudes. It is imperative that an effective dialogue regarding future development and application of the new biotechnology is established with the interested public. Wider involvement of the public in the development of the regulatory framework would be facilitated by the development of effective communication strategies, although there is a dearth of information at present as to the most effective way to develop such risk–benefit communication.

References

Barke, R. P. and Jenkins-Smith, H. C. (1993) Politics and scientific expertise: scientists, risk perception, and nuclear waste policy. *Risk Analysis*, 13 (4), 425–39.

Bauer, M. (1995) Resistance to new technology and its effects on nuclear power, information technology and biotechnology, in *Resistance to New Technology* M. Bauer (ed.), Cambridge University Press, Cambridge, pp. 1–45.

Buchmann, M. (1995) The impact of resistance to biotechnology in Switzerland: A sociological view of the recent referendum, in *Resistance to New Technology* M. Bauer (ed.), Cambridge University Press, Cambridge, pp. 189–208.

Cantley, M. (1987) Democracy and biotechnology: Popular attitudes, information, trust and public interest. *Swiss Biotech*, 5 (5), 5–15.

Capek, S. M. (1993) The 'environmental justice' frame: a conceptual discussion and an application, *Social Problems*, 40, 5–24.

Covello, V. T., von Winterfeldt, D. and Slovic, P. (1986) Risk communication: A review of the literature. *Risk Abstracts*, 3, 171–82.

De Flines, J. (1987) Publieke opinie essentieel voor biotechnologie, Interview. *De Ingenieur*, 6, 6–9.

Eagly, A. H. and Chaiken, S. (1993) *The Psychology of Attitudes*. Harcourt Brace Jovanovich, New York.

Eurobarometer 39.1 (1993) *Biotechnology and Genetic Engineering: What Europeans think about it in 1993*. INRA for the Commission of European Communities, Brussels.

Fischhoff, B. (1995) Risk perception and communication unplugged: Twenty years of process. *Risk Analysis*, 15 (2), 137–46.

Fischhoff, B., Bostrum, A. and Quadrel, M. J. (1993) Risk perception and communication. *Annual Review of Public Health*, 14, 183–203.

Fischhoff, B. *et al.* (1978) How safe is safe enough? A psychometric study of attitudes towards technological risks and benefits. *Policy Sciences*, 9, 127–52.

Flynn, J., Slovic, P. and Mertz, C. K. (1993) Decidedly different: Expert and public views of risks from a radioactive waste repository; *Risk Analysis*, 13, (6), 643–48.

Foreman, C. T. (1990) Food safety and quality for the consumer: policies and communication. In NABC report 4, *Animal Biotechnology: Opportunities and Challenges*. J. Fessenden Macdonald (ed.), National Agricultural Biotechnology Council, Ithaca, NY, pp. 121–6.

Frewer, L. J., Raats, M. M. and Shepherd, R. (1993/1994) Modelling the media: The transmission of risk information in the British press. *IMA Journal of Mathematics Applied in Business and Industry*, 5, 235–47.

Frewer, L. J . and Shepherd, R. (1994) Attributing information to different sources: Effects on the perceived qualities of the information, on the perceived relevance of the information, and on attitude formation. *Public Understanding of Science*, 3, (4), 385–401.

Frewer, L. J., Howard, C. and Shepherd, R. (1995) Genetic Engineering and Food: What determines Consumer Acceptance? *British Food Journal*, 97 (8), 31–7.

Frewer, L.J. and Shepherd, R. (1995) Ethical concerns and risk perceptions associated with different applications of genetic engineering: Interrelationships with the perceived need for regulation of the technology, *Agriculture and Human Values*, 12 (1) 48–57.

Frewer, L. J. *et al.* (1996a) Consumer concerns about food processing technologies: implications for effective communication, Proceedings, Joint annual meeting of Agriculture and Human Values Society, Association for the Study of Food and Society, and the International Food Choice Conference, St Louis, US, 6th–8th June 1996.

Frewer, L. J. *et al.* (1996b) What determines trust in information about food-related risks? Underlying psychological constructs, *Risk Analysis*, 16, 473–86.

Frewer, L. J. *et al.* (1997a) 'Objection' mapping in determining group and individual concerns regarding genetic engineering, *Agriculture and Human Values*, in press.

Frewer, L. J., *et al.* (1997b) Consumer attitudes towards different food-processing technologies used in cheese production – the influence of consumer benefit. *Food Quality and Preference*, 8, 1–10.

Frewer, L. J., Howard, C. and Shepherd, R. (1997) Public concerns about general and specific applications of genetic engineering: Risk, benefit and ethics, *Science, Technology and Human Values*, 22, 98–124.

Godown, R. D. (1987) The Science of Biotechnology, in *Public Perceptions of Biotechnology*, L. R. Batra and W. Klassen (eds), Agricultural Research Institute, Maryland.

Hagerstand, T. (1967) *Innovation Diffusion As A Spatial Process*. University of Chicago Press, Chicago.

Hamstra, A.M . (1991) *Biotechnology in Foodstuffs. Towards a Model of Consumer Acceptance, Research report 105*, SWOKA, The Hague, The Netherlands.

Hamstra, A. (1992) Consumer Research on biotechnology, in *Biotechnology In Public. A Review Of Recent Research*, J. Durant (ed.) *Science Museum*, London, pp. 42–51.

Heijs, W. J. M., Midden, C. J. H. and Drabbe, R. A. J. (1993) *Biotechnology: Attitudes and Influencing Factors*, Eindhoven University of Technology, Eindhoven.

Hoban, T. J. (1996a) Trends in consumer acceptance and awareness of biotechnology. *Journal of Food Distribution Research* 27 (1), 1–10.

Hoban, T.J. (1996b) How Japanese consumers view biotechnology. *Food Technology* July, 85–8.

Hoban, T. J. and Kendall, P. A. (1992) *Consumer Attitudes about the Use of Biotechnology in Agriculture and Food Production*, North Carolina State University: Raleigh, NC.

Hoban, T. J. and Kendall, P. (1993) Public perceptions of the benefits and risks of biotechnology, in *Agricultural Biotechnology: A Public Conversation about Risk*, J. Fessenden MacDonald (ed.), National Agricultural Biotechnology Council, Ithaca, New York, pp. 73–86.

Hovland, C. I., Janis, I. L., and Kelly, H. H. (1953) *Communication and Persuasion: Psychological Studies of Opinion Change*. Yale University Press, New Haven, CT.

Jasanoff, S. (1993) Bridging the two cultures of risk analysis, *Risk Analysis*, 13 (2), 123–9.

Jasanoff, S. (1995) Product, process, or programme: Three cultures and the regulation of biotechnology, in *Resistance to New Technology* M. Bauer (ed.), Cambridge University Press, Cambridge, pp. 335–56.

Joss, S. and Durant, J. (1995) The UK national consensus conference on plant biotechnology, *Public Understanding of Science*, 4, (2), 195–204.

Kasperson, R., Golding, D. and Tuler, S. (1992) Social distrust as a factor in siting hazardous facilities and communicating risks, *Journal of Social Issues*, 48, 161–87.

Kepplinger, H. M. (1995) Individual and Institutional impacts upon press coverage of sciences: The case of nuclear power and genetic engineering in Germany, in *Resistance to New Technology* M. Bauer (ed.), Cambridge University Press, Cambridge, pp. 357–78.

Knudsen, I. and Ovesen, L. (1994) Assessment of novel foods: A call for a new and broader GRAS concept. *Regulatory Toxicology and Pharmacology*, 21, 365–9.

Lieppe, M. R., Greenwald, A. G. and Baumgardner, M. H. (1982) Delayed persuasion as a consequence of associative interference: A context confusion effect. *Personality and Social Psychology*, 25, 65–74.

Liu, J. T. and Smith, V. K. (1990) Risk communication and attitude change: Taiwan's national debate over nuclear power. *Journal of Risk and Uncertainty* 3, 331–49.

Petersen, J. C. (ed.) (1984) *Citizen participation in Science Policy*, University of Massachussets Press: Amherst, MA.
Marlier, E. (1992) Eurobarometer 35.1, in *Biotechnology In Public. A Review Of Recent Research*, J. Durant (ed.) Science Museum, London, pp. 52–108.
Martin S. and Tait, J. (1993) *Release Of Genetically Modified Organisms: Public Attitudes And Understanding*. Open University, Centre for Technology Study, Milton Keynes.
Mazur, A. and Lee, J. (1993) Sounding the global alarm: Environmental issues in the US national news. *Social Studies of Science* 23, 681–720.
Mazur, A. (1987) Putting Radon on the public's risk agenda. *Science, Technology and Human Values* 12, 86–93.
McCombs, M. E. and Shaw, D. L. (1972) The agenda-setting function of the mass media. *Public Opinion Quarterly* 36, 176–87.
Michael, M. (1992) Lay discourses of science – Science in general, science in particular and self. *Science, Technology and Human Values*, 17, 313–33.
Michie, S. *et al.* (1995) A comparison of public and professionals' attitudes towards genetic developments, *Public Understanding of Science*. 4 (3), 243–55.
Mirowsky, J. and Ross, C. E. (1986) Social patterns of distress, *Annual Review of Sociology*. 12, 23–45.
Nelkin, D. (1995) Forms of Intrusion: Comparing resistance to information technology and biotechnology in the USA, in *Resistance to New Technology* M. Bauer (ed.), Cambridge University Press, Cambridge, pp. 379–92.
National Consensus Conference, Final report, (1994) *UK National Consensus Conference on Plant Biotechnology*, Science Museum, London.
Plough, A. and Krimsky, S. (1987) The emergence of risk communication studies: Social and political context. *Science, Technology and Human Values*, 12, 4–10.
Powell, D. A. and Griffiths, M. W. (1993). Public perceptions of agricultural biotechnology in Canada, Proceedings, Food Technology, August. pp. 14.
Ruhrmann, G. (1992) Genetic Engineering in the press: A review of research and results of a content analysis, in *Biotechnology in Public. A Review of Recent Research*. J. Durant (ed.), Science Museum for the European Federation of Biotechnology, London, pp. 169–201.
Sjoberg, L. and Drottz-Sjoberg, B. M. (1994) *Risk Perception Of Nuclear Waste: Experts And The Public. Report No. 16*. Stockholm School of Economics: Stockholm.
Slovic, P., Fischoff, B. and Lichtenstein, S. (1980) Risky assumptions, *Psychology Today*, 14, 44–88.
Slovic, P., Kraus, N. and Covello, V. (1990) What should we know about making risk comparisons? *Risk Analysis*, 10, 389–92.
Slovic, P. (1993) Perceived risk, trust and democracy, *Risk Analysis*, 13, (6), 675–82.
Smink, G. C. J. and Hamstra, A. M. (1996) *Informing Consumers About Foodstuffs Made With Genetic Engineering: A Constructive Contribution To The Issue*. Swoka, Leiden.
Sparks, P. and Shepherd, R. (1994) Public perceptions of the hazards associated with food production and food consumption: An empirical study, *Risk Analysis* 14, (5), 79–86.
Sparks, P., Shepherd, R. and Frewer, L. J. (1994) Gene technology, food production and public opinion: A UK study. *Agriculture and Human Values*, 11, (1), 19–28.
Stenholm, C. W. and Waggoner, D. B. (1992) Public Policy, in *Animal Biotechnology in the 1990's: Opportunities and Challenges*, J. Fessenden MacDonald (ed.), National Agricultural Biotechnology Council, Ithaca, NY, pp. 25–35.
Stewart, J., Kendall, E. and Coote, A. (1994) *Citizens' Juries*, Institute for Public Policy Research, London.
Straughan, R. (1992) *Ethics, Morality and Crop Biotechnology*, ICI Seeds, Fernhurst, Surrey, UK.
Tversky, A. and Kahneman, D. (1981) The Framing of decisions and the psychology of Choice, *Science*, 211, 453–8.
Tyler, T. R. and Griffin, E. (1991) The influence of decision makers goals on their concerns about procedural justice, *Journal of Applied Social Psychology* 21, 1629–58.
Vaughan, E. (1993) Individual and cultural differences in adaptation to environmental risks, *American Psychologist*, 48, 1–8.
Wynne, B. (1989) Sheep farming after Chernobyl: A case study in communicating scientific information, *Environment*, March, 10–15 and 33–39.
Zechendorf, (1994) What the public thinks about biotechnology, *Bio/Technology*, 12 September, 870–75.

3 Moral concerns and the educational function of ethics

ROGER STRAUGHAN

3.1 Introduction – why bother about ethics?

Modern food biotechnology is a highly technical subject, involving processes and products which stem from research in genetic engineering undertaken at the frontiers of science, as chapters in the latter part of this book will demonstrate. The can of genetically modified tomato paste, for example, sitting unobtrusively on the supermarket shelf, represents a great deal of innovative research and development. Why then does this specialized area of applied science need to be further complicated by introducing the extra dimension of *ethical* considerations? Ethics (in one sense of the word at least) is itself a complex philosophical activity requiring rigorous, analytical study. There is, therefore, an obvious danger that any investigation of the ethics of food biotechnology may become an abstruse, theoretical pastime of interest only to a few professional philosophers of science.

Fortunately, there is a much more down-to-earth approach which we can adopt here, which recognizes that there are compelling practical reasons why the ethics of food biotechnology can and should be a matter of concern to us all, rather than a subject of esoteric interest.

The first of these reasons is that *no* area of scientific or technological development can avoid some kind of ethical scrutiny. The fact that new technologies exist does not mean that they necessarily *ought* to be employed. The pursuit of new knowledge and techniques can never be given a total ethical *carte blanche*. Science cannot be pursued in a complete moral and ethical vacuum in any society that claims to be healthy and civilized; the universal condemnation of 'medical research' as pursued in Nazi Germany supports this view.

Moreover, it is probably unfair and presumptuous to assume that any scientist, technologist or industrialist *wants* to operate in such a vacuum. As rational human beings these people will normally have the same appreciation of the moral dimension of life as anyone else, and that dimension will span the areas of science, technology and industry. Moral and ethical questions, then, cannot be by-passed or ignored in biotechnology, any more than they can elsewhere.

The second, more pragmatic reason is that ethical considerations play an important part in influencing public attitudes and perceptions, as was shown in the previous chapter. Our attitudes are not, of course, wholly determined by our moral views alone, for other factors also play a part. Nevertheless, our moral views about any subject will clearly exert a powerful effect upon our perceptions of that subject and upon our attitudes towards it, for our moral beliefs play an integral part in how we see and interpret the world, which in turn helps to shape our choices and behaviour.

Concern is being increasingly expressed that the potential benefits of biotechnology may be lost if the new processes and products fail to gain 'consumer acceptance':

> Biotechnology appears to offer the best tools for achieving sustainable agricultural systems, while providing a safe, nutritious, abundant and affordable food supply. So, why the controversy? There is no simple answer to the question, but it is clear that social, moral and ethical issues, as well as scientific and technical matters, are at the heart of the debate.
>
> (Harlander, 1991)

There must, then, be a strong practical argument in favour of examining the ethical issues, not in order to try paternalistically to persuade the public that genetic engineering is really a good thing, but to raise the debate above the level of emotive propaganda and to encourage judgements to be made on a more rational and considered basis. If there is indeed a sound ethical case to be made in favour of food biotechnology, as its proponents claim, then its progress can only be aided by a detailed analysis of the concerns which have been voiced.

3.2 What are 'moral and ethical concerns'?

Before proceeding further, we need to clarify what precisely is meant by 'moral and ethical concerns'. The two adjectives are often used interchangeably in this context (which would make one of them redundant), but can any useful distinctions be drawn between them?

Certainly there is no single agreed way of using these terms (even among moral philosophers), but for the purposes of this chapter it may be helpful to distinguish between two levels of 'concern' which are frequently confused in debates about biotechnology.

3.2.1 'Moral' concerns

Most if not all human beings can be said to have moral views, beliefs and concerns, to the effect that certain things are right or wrong and certain actions ought or ought not to be performed. These views may refer to virtually any subject; a person may feel that it is wrong to swear, or to have extra-marital sexual relationships, or to smack children. Such 'moral concerns' may result from a lot of deliberation and reflection, or from very little; they may be firmly grounded in a consistent set of carefully considered rational principles, or they may not; their justification may have been consciously analysed, or it may not. We all probably hold some moral views almost unthinkingly, perhaps as a result of our upbringing. We may just 'feel' that certain things are right or wrong; we have a 'gut reaction' about them; and that may be the sum total of some people's 'morality'.

3.2.2 'Ethical' concerns

Ethics is a narrower concept than morality, and it can be used in several different, though related, senses. The most general of these:

> ... suggests a set of standards by which a particular group or community decides to regulate its behaviour – to distinguish what is legitimate or acceptable in pursuit of their aims from what is not. Hence we talk of 'business ethics' or 'medical ethics'.
>
> (Flew, 1979)

More technically, ethics can also refer to a particular branch of philosophy – moral philosophy. There is disagreement among philosophers about the precise scope, function and methodology of ethics, but this is not the place for a detailed discussion of such issues. For our present purposes we should note that one central task of ethics is usually taken to be a critical investigation of the fundamental principles and concepts that are used in moral debate. Ethics tries to analyse and clarify the arguments that are used when moral questions are discussed and to probe the justifications that are offered for moral claims. So ethics in this sense is a critical 'second-order' activity which puts our 'first-order' moral beliefs under the spotlight for scrutiny, but this is not an activity reserved solely for professional philosophers. We all engage in ethical thinking when we try to reflect in a detached way about the general reasons why we feel that something is morally right or wrong.

This distinction between 'moral' and 'ethical', therefore, shows how essential it is to 'unpack' the apparently straightforward statement that biotechnology is a source of moral and ethical concern; for to call something a *moral* concern does not necessarily mean that it is of much *ethical* significance. Many people in the past felt moral concern about old ladies who lived alone with black cats, but what ethical validity did such feelings have? Biotechnology may be rather a similar source of moral concern to some people today. If asked, people will express *moral* concern about genetic engineering (as they would have done about old ladies with black cats), but this does not tell us whether they have done any *ethical* thinking about the issues.

One fruitful way, then, of tackling the ethics of food biotechnology is to focus on those issues which appear to give rise to most moral concern. By looking at the ways in which these moral concerns are often expressed, we can try to probe the concepts and principles which underlie them and subject them to some ethical scrutiny. As one American commentator puts it:

> A significant part of the current debate can be traced to differences over moral principles. Also, unfortunately, there has been much unnecessary debate generated by careless moral reasoning and a failure to attend to the logical structure of some of the moral arguments that have been advanced.
>
> (Stich, 1989)

Moral disagreements, however, cannot be resolved by a simple appeal to 'the facts', for sets of facts whether about biotechnology or anything else can never *prove* something to be morally right or wrong. Nor can ethics offer conclusive, prescriptive *answers* about the rightness or wrongness of food biotechnology. One cannot prove that the hungry *ought* to be fed in the same way that one can prove that lack of food causes death. Moral judgements are *decisions*; they may be argued for or against, criticized or defended and shown to be more or less rational and informed, but their rightness or wrongness can never be comprehensively established. The purpose of this chapter, then, is more to pose questions than to answer them.

3.3 Why should there be moral concerns about food biotechnology?

It is not surprising on reflection that modern food biotechnology should cause many people concern, often of a moral kind. Firstly, food is immensely important to us for all kinds of reasons. As the Canadian writer, Margaret Visser (1993) comments:

> . . . people have always cared intensely about what they eat. People are incredibly fussy about their food. After all, food has to cross the thresholds of our mouths. The outside is taken in. For the whole of evolution, the process has been momentous. What is this foreign substance? How can I be really sure that I can safely ingest it? I need to eat it or I'll die, but if I eat the wrong thing, I could pay a heavy price. It could even kill me.

Anything that might be thought to threaten the purity of our food, therefore, is likely to arouse suspicion and righteous concern, as has been demonstrated by a number of recent 'food scares' in various countries. It is easy, then, for any interference with the fundamental genetic structure of our food to come to be seen as a threatening form of pollution and therefore morally wrong. To quote Margaret Visser again:

> The primary idea of pollution is a culinary one. Spoiled food is polluted. And in some cultures polluted people are untouchable just like the tainted food. There are people you cannot eat with. Polluted things are matter out of place.
>
> The modern, scientifically verifiable idea of pollution is not all that different, if you come to think of it. Oil in itself is fine, but oil in the sea is a menace. Images of sea birds covered with black slime are enough to make any civilized person shudder. The revulsion and the pity turns swiftly into moral outrage. It is our greed, our ignorance, our presumption – and our failure to control all of them – that has done this.
>
> Human beings have a strong sense that mixing the categories that nature has taken eons to distinguish cannot be wise. They instinctively feel that combining two living things that do not go together, that will not breed, for example, must be wrong. Horror movies dote on terrifying images of nature subverted.

Secondly and more generally, the whole image of genetic engineering can generate strong emotional and moral reactions. Many surveys have revealed public concern about the 'unnaturalness' of modern biotechnology and have shown it to be closely associated in many lay people's minds with that which is thought to be monstrous and frightening (Straughan, 1995). Fear of the unknown is probably an important factor here, though negative reactions of this kind are by no means limited to those who are ignorant or uninformed about the subject. In the UK, for example, the Prince of Wales (1995) in a well-publicized speech to an International Biodiversity Seminar in London, spoke of 'feeling profoundly apprehensive about many of the early signals from this brave new world, and the confidence – bordering on arrogance – with which it is promoted'. Some applications of genetic engineering, he claimed, were 'enough to send a cold chill down the spine'. Similar reservations have been eloquently expressed for many years by campaigners such as Jeremy Rifkin in the USA and more recently by various pressure groups in Germany, Holland and the UK.

3.4 Is modern biotechnology wrong in itself?

If we start to look in more detail at the main moral concerns which have been expressed, we find that the most fundamental of these stem from the nature of the technology itself. Thus, food biotechnology involving genetic engineering will be thought to be morally wrong by some, not because food products are specifically involved, but because of unease at the very idea of manipulating genetic material.

Why might genetic manipulation be thought to be wrong in itself? Most of the arguments here depend on the assumption that the technology is morally wrong because it is in some way 'unnatural'. They are in the main 'metaphysical' arguments in the sense that they incorporate a view of the world, mankind and nature which cannot be proved or disproved by pointing to sets of physical facts. Many, but by no means all, are also religious in that they depend upon beliefs about God and his relationship to the natural world. Many of the arguments are inter-related, though it is quite possible to support some while rejecting others.

The main strands of these arguments and the beliefs that underlie them can be briefly listed as follows:

1. Genetic manipulation involves 'playing God', by tinkering with the stuff of life. God, not man, is the owner and director of Nature (often personified in such arguments with a capital N). We are attempting a second Genesis, and before we displace the first Creator we should reflect whether we are qualified to do as well.
2. Genetic manipulation assumes a 'reductionist' view of life, encouraging us to adopt the chemist's perspective and look upon all forms of life as 'just DNA'.

   As Jeremy Rifkin puts it:

   > ... the important unit of life is no longer the organism, but rather the gene.... From this reductionist perspective, life is merely the aggregate representation of the chemicals that give rise to it and therefore (those adopting this perspective) see no ethical problem whatsoever in transferring one, five or a hundred genes from one species into the hereditary blueprint of another species. For they truly believe that they are only transferring chemicals coded in the genes and not anything unique to a specific animal. By this kind of reasoning, all of life becomes desacralized. All of life becomes reduced to a chemical level and becomes available for manipulation.

3. Genetic manipulation breaches barriers and boundaries between species which Nature has set up through the process of evolution to prevent genetic interactions between species, possibly for some overriding evolutionary reason . Alternatively, the 'creationist' view is that all existing species were created once and for all by God, and attempting to modify this arrangement constitutes a form of blasphemy. Both perspectives see species as 'sacred', either in evolutionary or creationist terms, and genetic manipulation as a violation of this 'sacredness'.
4. Genetic manipulation distorts mankind's relationship with the rest of nature. By engineering plants and animals for our own purposes, we come to assume that we own other life forms. The possibility of patenting new life forms encourages this arrogance.

To examine the above arguments and their underlying assumptions fully would require a whole book. All that can be offered here is a brief list of comments and considerations, some general and some specific, which may help us in judging the ethical validity of some of the above claims.

Firstly, there are obvious and notorious difficulties involved in trying to define what counts as 'natural' and 'unnatural'. What do these words mean when today we are offered 'natural' toothpaste, margarine and a host of other products in which man (and woman) seem to have had much more of a hand than Nature? How do we decide whether or not it is 'natural', for example, for women to prefer child-rearing to a career, for boys to play with dolls, for huskies to live as household pets, or for roses to be blue? How can we ever be sure what, if anything, 'Nature' prescribes?

Moreover, even if we could be sure about this, it would not have any direct ethical relevance – something is not made automatically good or right just because it is 'natural'. Some of the most likely examples of 'natural' human characteristics would include such dispositions as jealousy, aggression, insecurity, possessiveness and self-centredness, but what is morally desirable about these? Many substances and events in Nature cause harm and suffering; some of them we call 'natural disasters'. Just because something happens in Nature does not tell us what we morally ought to do about it. A value judgment always has to be made *about* the so-called 'facts of nature' – even when we think we can identify them. These logical points highlight the dangers of sweeping statements about the 'unnatural' aspects of genetic manipulation and of assumptions that moral implications follow clearly from them.

Secondly, the concept of 'natural species barriers' is particularly obscure. Much controversy among American scientists in the 1970s centred on whether or not such barriers existed, and – an example is given below – it was found that clear-cut answers were not available:

> . . . the issue whether there are genetic barriers separating species is a complicated affair. Partly, this is because most discussions have not clearly explicated the meaning of species barriers. . . . Inquiry into these matters must consider three questions: (1) What types of barriers divide species? (2) What forms of genetic exchange exist apart from what can be achieved through technology? (3) Which barriers, if any, can be uniquely breached by DNA technology?
>
> (Krimsky, 1982)

This is not the place for a detailed, technical discussion of the issue. We must merely again note the danger of simplistic assumptions and realize that, as in (1) above, even if a natural barrier exists, that does not prove that it is morally wrong to cross it.

Thirdly, genetic manipulation is incompatible only with some, and by no means all, sets of religious beliefs. It is, for example, quite possible to believe in a God who does not see species as particularly 'sacred', but as provisional within the evolutionary process which he has initiated and of which innovative man is a product. Complementary to this view is the belief that man should 'use his talents' (as recommended in the parable) by developing to the full all the knowledge and skill which is available to him, and act as 'co-creator' with God. Genetic manipulation is thus not necessarily objectionable to religious believers, and in fact a wide diversity of views exists both among different religions and within a single religion such as Christianity (Reiss and Straughan, 1996).

Fourthly, man has always 'played God' in his attempts to 'improve' breeds of animal and strains of plants for his own ends – every dog show or seed catalogue provides evidence of this. Genetic manipulation is seen by many, therefore, as merely a more efficient and speedy method of doing what man has always tried to do. The key question here is whether or not genetic manipulation possesses features which make it ethically distinctive from traditional methods of selective breeding.

The above points are not intended to 'answer' all the moral and religious arguments about genetic manipulation. They do suggest, however, that some of the arguments rest upon shaky foundations and obscure concepts, often presented in an emotive form. Nevertheless, although the cumulative force of these arguments is by no means irresistible, it is clearly possible to hold moral, religious or metaphysical beliefs which require the rejection of genetic manipulation.

In view of this potential for public concern, it is not surprising that a number of government committees have been set up in various countries to investigate and report on the issue. In the UK, for example, the Polkinghorne Committee on the Ethics of Genetic Modification and Food Use (1993) was asked 'to consider future trends in the production of transgenic organisms; to consider the moral and ethical concerns (other than those related to food safety) that may arise from the use of food products derived from production programmes involving such organisms; and to make recommendations'. In its conclusions the Committee recognized:

> ... that many groups or individuals within the population object on ethical grounds to the consumption of organisms containing copy genes of human origin. We therefore recommend that food products containing such organisms should be labelled accordingly to allow consumers to exercise choice.
>
> We recognise that some groups or individuals within the population object on ethical grounds to the consumption of organisms containing copy genes from animals which are the subject of dietary restrictions for their religion. We recognise that similar ethical objections would be held by certain vegetarians in relation to any copy gene of animal origin incorporated in a plant. We recommend that appropriate food products should be labelled accordingly to allow these groups to make an informed choice.

Another example of a government report which took a broader view of the possible ethical problems than did Polkinghorne comes from Sweden (Swedish Government Committee, SOU, 1992). This questioned whether nature has an inherent value and whether human beings have the right to change nature. It concluded as follows:

> The Committee is of the opinion that nature does have an inherent value. But the Committee also believes that humans, under certain conditions, have the right to modify nature and what lives in nature. The Committee is thus of the opinion that the overriding principle should be that *humans do have the right to modify plants, animals and micro-organisms in order to improve their living conditions if the modifications can be accomplished with due respect to the doctrine of nature conservation and without harming humans or animals*. The right of humans in accordance with this principle must nevertheless be linked to a moral responsibility. The natural organisms have resident inherent values which must be respected. According to the Committee, humans are part of a meaningful whole and also have a duty to treat animals well.
>
> If human actions abide by these ethical guidelines, the Committee is of the opinion that it may be defensible to modify heredity properties of animals, plants and micro-organisms. Humans have been making such changes for a very long period of time in breeding animals and plants. By means of genetic engineering, however, barriers between species can be bridged, which makes it important to adopt an ethical stance and to consider whether or not the intervention can take place without damage to individuals and ecosystems. Nevertheless, according to the Committee caution should be the yardstick when interfering with nature. As a precaution all serious interference with nature should be avoided. Such interference as is made should be reversible.

Such references to possible damage and to the need for caution take us beyond concerns that modern biotechnology may be wrong in itself by focusing on its possible *consequences*, something to which we must now turn.

3.5 Do the possible consequences of modern biotechnology warrant moral concern?

Concerns about possible risks are not unconnected with fears about 'unnaturalness', for whatever is deemed unnatural is often also deemed unsafe, as countless horror movies demonstrate. However, an important distinction can be drawn between concerns that modern biotechnology in all its applications may be wrong *in itself* and concerns that some of its applications may be wrong because of possible *consequences*. If any activity or process is thought wrong in itself, its consequences become irrelevant, for nothing can affect its intrinsic wrongness. But if an activity or process is seen as morally neutral with a potential for good or ill, then a consideration of the possible consequences will play a major part in any ethical assessment (Reiss and Straughan, 1996).

First, we need to ask whether safety and risk are really ethical or moral matters. In one sense, safety does not appear to be a moral or ethical question. The safety of a product, process or activity is, at least on the face of it, an empirical matter to be determined by experiment and experience.

Whether or not a toadstool is safe to eat, for instance, is not an ethical question. Yet questions about safety can be closely related to and can indeed raise ethical questions. To develop the example just given, if it is known that poisonous toadstools grow in profusion in a particular public area, what steps if any should be taken and by whom to prevent people from eating them? Questions about safety, then, raise further ethical questions about responsibility and accountability and about acceptable or justifiable levels of risk. The Swedish Government report referred to above supports this view:

> The Committee takes the view that the assessment of risks should be part of the ethical analysis, since it is ethically false to base a decision on poor foundations if the decision can be postponed until the foundations have improved. It is also ethically unacceptable to assert that the foundations for a decision are better than they are. Thus the open presentation of facts and viewpoints is important.
>
> (Swedish Government Committee, 1992)

These general considerations of responsibility, accountability, acceptability and justifiability are all highly relevant to the genetic manipulation of plants, animals and microbes for the purposes of food production.

The main fears centre upon the deliberate release of genetically manipulated organisms into the environment and the possible harmful consequences of such procedures. Such consequences might include, it has been suggested, global drought caused by the genetically modified 'ice-minus' organism (intended to protect crops against frost) and the spread of indestructible weeds, resistant to pests and herbicides. Further fears have been voiced that crop biotechnology might lead to a loss of genetic diversity: a reduced number of 'supercrops' might prove to be less resilient and so more vulnerable to various forms of attack in the future.

The risks envisaged here are clearly of such a catastrophic nature that no one (with the exception of the archetypal 'mad scientist' of sensational fiction) would feel justified in turning a blind eye to them. So can we cut short our ethical investigation at this point by accepting that such risk-taking is irresponsible and unjustifiable on the basis of a moral principle to the effect that any activity which could lead to catastrophic consequences ought not to be undertaken? Unfortunately, this simple and apparently responsible conclusion becomes less convincing when we look more closely at the moral principle on which it depends.

The problem is that in theory *any* activity could lead to catastrophic consequences. Eating beefburgers in the 1980s in the UK was not generally thought of as a hazardous activity, but the possibility of that activity resulting in catastrophic consequences is now considered not insignificant by many scientists. Nothing can ever be guaranteed or proved to be 100% safe.

Scientists cannot 'prove' by empirical investigations that one experiment or class of experiments is more hazardous than another without undertaking experiments that are in fact hazardous. Nor can they 'prove' by tests and experiments that a particular event will never happen in the future. We can talk only in terms of apparent probabilities. Godown (1987) illustrates this point forcefully:

> Can science tell us for instance what will be the result of creating and releasing a novel organism from which a single gene has been deleted? Could it ever be flatly stated on the basis of scientifically established facts that there is no possibility of anything going wrong when a genetically engineered organism is deliberately released? The answer is obviously no, and I am willing to be quoted. One cannot prove a universal negative and it is silly to try.

Risks, then, are unavoidably involved in any activity, but the basic ethical question here is whether irresponsible and unjustifiable risks are being taken. Although this is not a question to which a clear-cut yes-or-no answer can be given, some weight must be given to the following considerations:

1. The possible harmful effects seem to be entirely speculative; no instances have occurred in practice. Unlike most new technologies in their early stages of development, modern biotechnology has so far proved to be remarkably safe.
2. Many scientists claim that this clean record indicates that modern biotechnology, far from being peculiarly risky, is in fact peculiarly safe – safer in fact than the relatively indiscriminate genetic exchanges that occur in traditional selective breeding, because the best a genetic engineer can do is add one gene, or at most a few genes, to tens of thousands of genes in an organism's chromosome.
3. Stringent regulations have been introduced in those countries where biotechnological developments are taking place. These are described elsewhere in this book and while it can always be argued that regulations are not strict enough, there seems little scope for irresponsible risk-taking of a Frankensteinian kind under the existing framework.
4. Excessive caution does not necessarily remove the risk of future catastrophes. By banning research and development in any new technology that is thought to involve risks, we may run the further risk of failing to produce an innovation which will be desperately needed in some future, unforeseen crisis. The history of science has proved to be highly unpredictable, and there can be no guarantee that 'playing safe' by abandoning research and development in biotechnology will not deny us a technique or product which may *prevent* an environmental disaster in 50 years' time.

The above considerations do not of course 'prove' the safety of food biotechnology (already shown to be an impossible task), but they do at least suggest that the case against it on the grounds of irresponsible risk-taking is far from overwhelming. Nevertheless, as it is logically impossible to guarantee the total safety of the technology, a value judgement has to be made about the acceptability and justifiability of the possible risks, and that judgement will involve the weighing of potential risks against potential benefits. As Stich (1989) emphasizes in his careful analysis: 'The question of how risks and benefits are to be weighed against each other . . . is the really crucial moral question raised by recombinant DNA research.'

3.6 Who will benefit?

The potential benefits of food biotechnology are described in detail elsewhere in this book and will not be repeated here. But who will reap these benefits, and will they be outnumbered by those who may suffer significant harm as a result of the new technology? Are there further moral concerns to be considered here?

The main worries appear to centre upon the economic vulnerability of poorer farmers and poorer countries to some of the possible effects of the new technology, and upon the economic disadvantage they are thought likely to suffer. Crop biotechnology will be expensive, it is claimed, and will favour large-scale, capital-intensive styles of agriculture; poor farmers will not be able to afford the new products and will lose out in the increased competition that will result. On a broader scale whole economies as well as individual farmers could be threatened, and the economic gap between the 'developing' and the 'developed' world could widen yet further.

It is particularly difficult to make a balanced ethical assessment here, as the issues are complicated by a variety of social, economic, political and agricultural factors about which genuine disagreement exists among experts and which cannot be explored in any great detail here.

Clearly, many of the developments in food biotechnology have great potential for increasing human welfare. In particular, in that problem area which many see as generating the most urgent moral imperatives of this generation – the equitable provision of food supplies throughout the world – many genetic manipulation techniques are specifically designed to increase dramatically the quantity and availability of food supplies by overcoming various geographical and climatic obstacles. If genetic manipulation can indeed help to alleviate the problems of world food supplies and eradicate hunger and starvation, this gives it an enormous potential 'moral plus'.

On the other hand, it can be argued that hunger and starvation are caused principally by political and economic factors rather than by direct food shortages, and that modern biotechnology is more likely to worsen rather than improve the lot of the poor and the deprived in various ways.

Trying to weigh the likely costs and benefits here and to arrive at a reasonable ethical judgement will require considerable technical expertise and accurate prediction, rather than the emotive rhetoric which frequently characterizes both sides of this debate. In general terms, however, we should bear in mind that all new technologies inevitably have far-reaching socio-economic effects. Food biotechnology cannot, then, be singled out as the sole target for moral censure on these grounds, any more than can information technology or the development of the steam engine, which caused some workers to lose their jobs and suffer economic hardship. Also, new technologies tend to benefit *initially* the 'developed' countries, because they have more resources and expertise available for research and development, though clearly there is a moral obligation upon those benefiting from a new technology to extend those benefits as widely and as speedily as is practicable.

One final moral concern relating to the potential costs and benefits of food biotechnology is that of animal welfare. There is considerable evidence that public attitudes are more negative generally towards animal applications of food biotechnology than towards plant or microbial applications, and one obvious reason for this is that animals are seen as sentient beings, capable of suffering and of having needs and interests in a way that plants and microbes are not. A number of animal welfare organizations have voiced objections to genetic modification, and it is certainly possible to point to examples of animals suffering as a result of the technology: early genetic experiments on pigs in the USA produced animals which exhibited a number of serious physical handicaps. However, public sensitivity to animal welfare issues makes it unlikely that such results would ever be acceptable, and it can be argued that biotechnology is more likely to reduce rather than increase animal suffering, by improving resistance to disease.

The ethical issues here though are more complicated than this. Even if it could be shown that animals were more likely to benefit than suffer from modern biotechnology in terms of reduced physical pain and discomfort, deeper problems remain about what it is right and wrong to do to animals. Would it be justifiable, for example, to engineer farm animals to make them more tolerant of stressful conditions arising from modern methods of intensive farming? Should we try to change the 'natural' dispositions and behaviour of animals in order to make them more productive – for example, by preventing turkeys from becoming broody (Reiss and Straughan, 1996)? Again, the complexity of these questions makes it impossible to probe them in detail here, but at least it is plain that there can be no simple, clear-cut answers.

3.7 Conclusions

This chapter has reviewed a wide variety of moral concerns about food biotechnology, and has tried to show how ethics can help to clarify exactly what is at stake and how we might set about making a reasoned decision about the issues. It must be emphasized, however, that this procedure cannot result in conclusive *proof* that food biotechnology in general or any particular application of it is morally right or wrong.

As such proof is impossible and as genuine moral decisions have always to be made *by* individuals or groups rather than *for* them, it is unjustifiable for politicians or philosophers or food producers or pressure groups to try to persuade the public that food biotechnology conclusively is morally right or wrong. Individual consumers will have to form their own judgements about this, and to help them do this a carefully planned education programme is called for, if irrationally polarized attitudes are to be avoided. This programme should not take the form of a propaganda exercise motivated by commercial interests; it should offer the public the basic facts about food biotechnology and should raise awareness both of its likely benefits and of the possible areas of concern with which this chapter has dealt.

Acknowledgement

Some of the material in this chapter is drawn from two independent reports by the author for the National Consumer Council, London, and for ICI Seeds (Zeneca).

References

Flew, A. (ed.) (1979) *A Dictionary of Philosophy*, Pan Books, London.
Godown, R. D. (1987) The science of biotechnology, in *Public Perceptions of Biotechnology* (eds L. R. Batra and W. Klassen) Agricultural Research Institute, Maryland, pp. 21–35.
Harlander, S. K. (1991) Communication between scientists and consumers. *Outlook on Agriculture*, 20.2. pp. 73–7.
Krimsky, S. (1982) *Genetic Alchemy*, MIT Press, Cambridge, Massachusetts.
Polkinghorne Report (1993), *Report of the Committee on the Ethics of Genetic Modification and Food Use*, HMSO, London.
Prince of Wales (1995) Speech at Lancaster House prior to International Biodiversity Seminar.
Reiss, M. and Straughan, R. (1996) *Improving Nature? The Science and Ethics of Genetic Engineering*, Cambridge University Press, Cambridge.
Rifkin, J. (1985) *Declaration of a Heretic*, Routledge & Kegan Paul, London.
Stich, S. P. (1989) The recombinant DNA debate, in *Philosophy of Biology* (ed. M. Ruse), Macmillan, New York, pp. 229–43.
Straughan, R. (1995) Monsters and morality, in *Genethics* (eds. H. P. Bernhard and C. Cookson), Ciba Chiron, Basle, pp. 28–31.
Swedish Government Committee (SOU) (1992) *Genetic Engineering – a Challenge*, Graphic Systems AB, Göteborg.
Visser, M. (1993) Thinking about food, Proceedings of Conference on *Symbol, Substance, Science – The Societal Issues of Food Biotechnology*, North Carolina Biotechnology Center, Research Triangle Park, North Carolina, pp. 5–13.

4 Worldwide regulatory issues: legislation and labelling

NICK TOMLINSON*

4.1 Introduction

The first genetically modified organism to be permitted for food use was a baker's yeast approved in the United Kingdom in 1990. The yeast which had been modified to improve dough quality (see also Chapter 8) was cleared on the advice of the United Kingdom's Advisory Committee on Novel Foods and Processes (ACNFP).

A large range of crops, yeasts and enzymes have now been produced using genetic modification techniques, although most of these products have yet to reach the stage of commercial development. It is clear that the use of such techniques is going to increase in importance in forthcoming decades. In future, genetically modified (GM) foods could offer consumers many benefits including improved choice, quality, flavour and keeping qualities, at lower prices. Equally, through the use of genetic modification techniques, it might be possible for crop yields to keep pace with the projected growth in the world population.

Many countries around the world are keen to benefit from modern biotechnology. At the same time there is general agreement on the need for appropriate safeguards to protect human health and the environment. Although a variety of regulatory approaches have been adopted around the world, they have many elements in common. The controls have been put in place, not because health or environmental problems have been identified, but because of a lack of familiarity with the behaviour of genetically modified organisms (GMOs). Controls were introduced before products reached the market in any significant quantities and these controls have resulted in a very good safety record for genetically modified foods.

Since many foods are traded internationally, companies developing genetically modified foods have to gain approval for their products in many countries. Looking to the future, it is inevitable that there will be increasing pressure to harmonise controls on GM foods. Already food safety

* The opinions stated are those of the author – they are not official statements of the organisation for which the author works.

assessment procedures in many countries are based on the concept of substantial equivalence developed by the OECD and subsequently refined by the World Health Organisation (OECD, 1993; FAO/WHO, 1996). Substantial equivalence is a comparative process in which a GM food is compared with a conventional counterpart. If differences are identified, safety assessment is focused on these differences.

4.2 Existing legislation

4.2.1 United Kingdom

The United Kingdom was one of the first countries in the world to introduce controls on GMOs back in 1978. These followed a precautionary approach because of the recognised lack of familiarity with the behaviour of GMOs at that time and the need to provide for safety. This precautionary principle has continued even though many of the earlier concerns have not materialised as our use and knowledge of the technology has progressed. Nevertheless, given the pace of its development, it remains important to ensure that no new hazards are created.

4.2.1.1 Environmental legislation. Work with GMOs in containment is covered by the Genetically Modified Organism (Contained Use) Regulations 1992 and 1995 made under the Health and Saftey at Work Act 1972. The Regulations implement the EC (European Commission) Directive 90/219/EEC on the Contained Use of GMOs. The EC Directive only applies to work with genetically modified microorganisms. However, the UK Regulations also apply to work with plants and animals.

Under the Contained Use Regulations those working with genetically modified organisms have to notify the Health and Safety Executive (HSE) and carry out an assessment of the risks that arise from the work that they are doing. If there is a risk of harm to humans or the environment, they must take the necessary measures to manage the risk. The focus of such a risk assessment is not so much the modification process itself but the organism that is being modified and the effect of the modification; there is clearly a greater level of risk arising from a modified human pathogen than a modified yeast. For the more hazardous microorganisms, the Health and Safety Executive, in consultation with other interested Government Departments, has to give a specific consent. In other cases there is a requirement to notify the HSE of proposed work at specific premises. All the premises notified to the Health and Safety Executive are subject to inspection by specialist Health and Safety Inspectors. In carrying out their responsibilities, the Executive are assisted by an independent advisory committee of scientific experts, the Advisory Committee on Genetic Modification.

The release of GMOs into the environment on an experimental basis is covered by the Genetically Modified Organisms (Deliberate Release) Regulations 1992 and 1995 made under the Environment Protection Act. The regulations implement the EC Directive 90/220/EEC on the deliberate release of GMOs. An experimental release, such as a field trial, cannot go ahead without a consent from the Secretary of State for the Environment. Where the Minister of Agriculture has an interest, as he has in all agricultural field trials, for instance, he acts jointly with the Secretary of State. In applying for a consent, an applicant is required to provide a considerable volume of data and also a detailed assessment of the risk of harm to human health and the environment. This risk assessment has to demonstrate that there is no significant risk arising from the release. If a risk is identified, or there is uncertainty as to whether such a risk will arise, the applicant may propose measures to manage and nullify such a risk. The applications and the risk assessments are scrutinised by the Advisory Committee on Releases into the Environment (ACRE), a group of independent experts who advise Ministers whether consent should be given, and whether any extra conditions should be imposed. All releases are advertised locally and the details are made available on a Public Register. Release sites are subject to inspection by the Health and Safety Inspectorate working for the Department of the Environment and those making the release are required to report any incidents that may occur during and following the completion of the release.

4.2.1.2 Food safety. In the UK, the safety assessment of all novel foods, including GM foods, has been undertaken by the ACNFP (Advisory Committee on Novel Foods and Processes), chaired by Professor D. C. Burke, since its formation in 1988. The ACNFP has operated a system based on a voluntary arrangement with the food industry which dates back to 1980. The 1990 Food Safety Act contained powers to put the ACNFP on a statutory basis. However, these were held in abeyance pending negotiations on an EC Novel Foods Regulation. The ACNFP's terms of reference are 'to advise Health and Agriculture Ministers of Great Britain and the Heads of the Departments of Health and Social Services and Agriculture for Northern Ireland on any matters relating to the irradiation of food or to the manufacture of novel foods or foods produced by novel processes having regard where appropriate to the views of relevant expert bodies'. In practice, the Committee advises Ministers to accept or reject a novel food on the basis of a rigorous safety assessment. The ACNFP developed a decision-tree-based approach, which was published in 1991 (Department of Health, 1991). This structured approach was refined in 1994 (ACNFP, 1994). This decision tree is used to identify the information requirements required for a particular type of novel food or process. The ACNFP's approach to assessing the safety of GM foods is based upon the concept of substantial equivalence. In assessing the safety of a novel food, the ACNFP can draw on expertise from a wide range of expert committees including the Department of Health's Committee on Toxicity (COT) and the Committee on Medical Aspects (COMA).

In the early 1990s, the ACNFP decided to address the wider issues relating to the food use of animals from genetic modification programmes, following receipt of a request to assess the safety of sheep modified to carry a human gene coding for the blood clotting protein Factor IX. The ACNFP recognised that there were ethical issues that went wider than those of safety. As a result, in 1992, the Minister of Agriculture appointed a Committee under the chairmanship of the Reverend Dr J. C. Polkinghorne to 'consider future trends in the production of transgenic organisms; to consider the moral and ethical concerns (other than those related to food safety) that may arise from the use of food products derived from production programmes involving such organisms; and to make recommendations' (Ministry of Agriculture, 1993). The Committee consulted widely with religious groups, consumer organisations and animal welfare groups. They found no overriding ethical objection requiring an absolute prohibition of the use of genetically modified organisms for food use. The Committee recognised that a number of individuals may object to consuming genetically modified food containing copy genes of animal or human origin. Consequently they recommended that foods containing such copy genes should be labelled accordingly. They also recommended that the use of ethically sensitive genes should be discouraged where alternatives were available.

4.2.1.3 Labelling. In the UK, the labelling of GM foods is considered by the independent Food Advisory Committee (FAC). The main food labelling legislation is the Food Labelling Regulations 1996, made under the Food Safety Act. These implement the EC Food Labelling Directive 79/112/EEC. The FAC first issued labelling guidelines for GM foods in 1993. These took account of two public consultation exercises as well as the views of the Polkinghorne Committee. The FAC concluded that only two main criteria should be used to determine which GM foods should be labelled, namely the source of the inserted gene, and whether the inserted genes are still present after processing. The FAC recommended that it would be unrealistic to label every food whose production has involved genetic modification. It did, however, accept that there should be provision for choice in relation to those foods which raise real concerns for a significant proportion of the population. It therefore proposed that a GM food should be labelled if it:

- contains a copy gene originally derived from a human;
- contains a copy gene originally derived from an animal which is the subject of religious dietary restrictions; or
- is plant or microbial material containing a copy gene originally derived from an animal.

These rules did not apply if the inserted copy gene had been destroyed by processing and was not, therefore, present in the food.

The FAC has been following developments on the EC Novel Foods Regulation very closely. As soon as the Council of Ministers agreed a common position on the Regulation, the FAC updated their labelling guidelines to reflect the provisions in the common position. The final regulation was agreed in December 1996 and the FAC immediately agreed to use the regulation as the basis for all future decisions on labelling GM foods, even though the regulation did not come into force until May 1997. The FAC recognises that many consumers are unfamiliar with genetically modified foods and has stressed that, particularly when labelling is not a condition of approval, manufacturers and retailers should be encouraged to provide additional information in response to public interest. This includes additional labelling information where appropriate. In response, a number of retailers and food organisations have produced an extensive array of information leaflets and customer care lines.

4.2.2 The European Union

4.2.2.1 Environmental legislation. In 1990 the Council of the European Communities adopted Directive 90/220/EEC on the deliberate release of genetically modified organisms. The Directive had two main objectives: to protect human health and the environment and to provide a harmonised regulatory framework covering all intentional releases of GMOs into the environment within the EC. The Directive covers release for experimental purposes and the commercial sale of GMOs. The Directive defines a GMO as 'an organism in which the genetic material has been altered in a way that does not occur naturally by mating and/or natural recombination'.

Under the Directive, all releases to the environment, whether for experimental purposes or for marketing, must be approved in advance. The person making an application for a consent to release a GMO is responsible for providing the required information. In addition, an applicant is required to provide a statement assessing the risks that the GMO poses to human health and the environment. This risk assessment is then evaluated by the competent authority to which the application is made.

In the case of experimental releases, a decision will be given within 90 days of an application being submitted to the competent authority in the EC Member State where it is intended to release the GMO, although the clock can be stopped if there is a need for further information. A summary of the information, together with the decision, are circulated to other states of the European Community. They cannot intervene in the decision but they may comment, for example, if they have particular knowledge of the intended release or particular expertise. In the light of experience gained, it is possible for a competent authority to introduce simplified procedures;

for example, in the case of a release of more than one GMO of the same species at several locations. In order to enable informed decisions to be made on future applications, an applicant is required to provide details of any relevant information obtained from the release that relates to risks to human health or the environment.

In the case of a release for marketing, the initial procedure is as for experimental releases. However, once the initial competent authority has reached a favourable opinion, this is communicated to the other 14 EC Member States who have 60 days in which to review the application and ask for further information or raise objections. If no objections are raised, the initial competent authority can issue a consent which is binding on the EC and the GMO can be marketed in all Member States in accordance with the marketing consent. However if, as has been the case with all the marketing applications submitted in the EC to date, objections are raised and these cannot be resolved between the Member States concerned, a vote is taken and the result of this is binding on all Member States. Details of GM crops for which a marketing consent has been issued in the European Community is given in Table 4.1.

Directive 90/220/EEC contains provisions for Member States to provide information, as appropriate, to the public and other groups. Once a marketing consent has been issued, no Member State can take action to prohibit or restrict the release of that GMO except where a Member State has new information that calls into question the basis upon which the marketing consent was issued, in which case they may provisionally restrict the sale of the GMO within their borders. Within a three-month period the Community has to reach a decision on the validity of the restriction. The Directive also allows for the marketing of certain GMOs to be exempted from its scope provided that marketing is governed by EC product legislation which contain provisions requiring an environmental risk assessment comparable to that required under the Directive. This will be the case with the implementation of the Novel Foods Regulation which contains provisions for the assessment of GMOs used for food.

Table 4.1 Genetically modified crops approved for sale in the European Community as of March 1997

Crop	Trait	Company
Chicory (non-food use)	Herbicide tolerance	Beijo Zaden BV
Maize	Insect resistance	Ciba Geigy
Oilseed rape (non-food use)	Herbicide tolerance	Plant Genetic Systems
Soya beans	Herbicide tolerance	Monsanto
Tobacco	Herbicide tolerance	SEITA

4.2.2.2 Food safety. In the European Community, the safety of genetically modified food is controlled by Regulation (EC) No. 258/97 concerning novel foods and novel food ingredients (1997). The regulation, which came into effect on 15 May 1997, introduced a mandatory pre-market safety assessment procedure for all novel foods. The regulation defines a novel food as a food or food ingredient which has not been used for human consumption to a significant degree within the Community and which falls under one of the following categories:

(a) food and food ingredients containing or consisting of genetically modified organisms within the meaning of Directive 90/220/EEC;
(b) foods and food ingredients produced from, but not containing, genetically modified organisms;
(c) foods and food ingredients with a new or intentionally modified primary molecular structure;
(d) foods and food ingredients consisting of or isolated from microorganisms, fungi or algae;
(e) foods and food ingredients consisting of or isolated from plants and food ingredients isolated from animals, except for foods and food ingredients obtained by traditional propagating or breeding practices and having a history of safe food use;
(f) foods and food ingredients to which has been applied a production process not currently used, where that process gives rise to significant changes in the composition or structure of the foods or food ingredients which affect their nutritional value, metabolism or level of undesirable substances.

The regulation does not cover novel food additives, flavourings or extraction solvents since these are covered by existing Community legislation. However, the Commission is required to ensure that safety levels applied to these groups of products correspond to the safety levels provided for by the Novel Foods Regulation. Before a novel food can be approved under the Regulation, it must satisfy three criteria. These are that the novel food must not: present a danger for the consumer, mislead the consumer, or differ from foods which they are intended to replace, to such an extent that their normal consumption would be nutritionally disadvantageous for the consumer. The regulation contains two main procedures for controlling novel foods. In the case of foods in categories (b), (d) and (e) that are considered to be substantially equivalent to existing foods, based either on the opinion of the assessment body in one Member State or on the basis of generally recognised scientific evidence, a notification procedure operates. For such products an applicant is required to inform the Commission when the product is first placed on the market. For all other novel foods, a full assessment procedure will be applied.

The assessment procedure is decentralised to Member States. A company wishing to market a novel food would apply to the Member State where they first intend to market the product. After an initial check to ensure that the application is complete, that Member State has 90 days in which to carry out an initial safety assessment and forward their opinion to the Commission. In the UK, applications are made to the Ministry of Agriculture, Fisheries and Food with the ACNFP assessing the safety of the product and the FAC considering the labelling of the product. The initial opinion can be either that the product should be approved or that a more detailed consideration is required. The Commission then forwards this initial opinion to the remaining 14 Member States who have 60 days to accept or object to the opinion. If no objections are raised, the applicant is informed that the product can be placed on the market. If one of more Member States raises an objection, the application is considered by the Standing Committee for Food, where a decision will be taken by qualified majority rating. The Standing Committee will seek the advice of the Scientific Committee for Food (SCF) on any issues relating to public health. To ensure that all Member States follow a similar approach to the safety assessment of novel foods, the SCF has recently produced a detailed set of guidelines setting out the type of information that would be expected to support an application for approval of a novel food.

The regulation is designed to follow the principle of 'one door one key' under which a company wishing to market a food that consists of a GMO can get all necessary approvals by making a single application. Consequently for such GMOs, the Regulation replaces the marketing consent provisions of the Deliberate Release Directive 90/220. For foods that fall into category (a) of the regulation, there is a requirement to assess the environmental risk from the product. This environmental risk assessment forms part of the initial assessment. In the UK, this will be considered by the Advisory Committee on Releases to the Environment. The various steps involved in dealing with an application for a food in category (a) are set out in Figure 4.1.

4.2.2.3 Labelling. Until the Novel Foods Regulation came into effect, foods containing or consisting of GMOs were authorised for marketing under Directive 90/220/EEC. This Directive which originally contained limited provisions for labelling, was recently amended to require the labelling of products containing or consisting of GMOs. The Novel Foods Regulations has introduced extensive labelling provisions which will apply in addition to existing community legislation on food labelling. All foods which contain or consist of GMOs within the meaning of Directive 90/220/EEC require labelling under the regulation. However, the regulation recognises that in some cases it may not be possible to segregate foods which contain genetically modified and conventional produce. In such circumstances, the

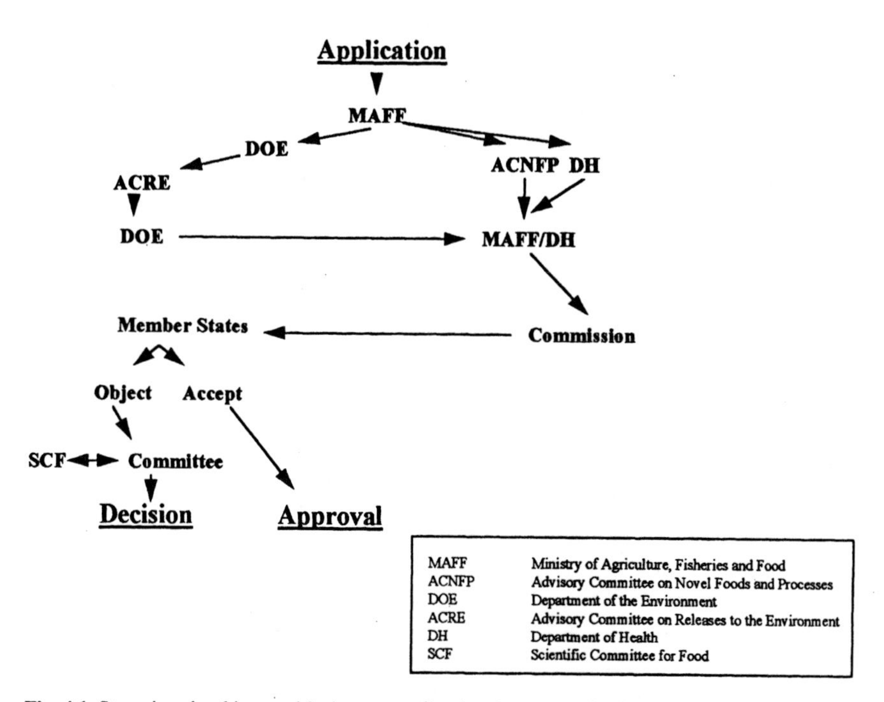

Fig. 4.1 Steps involved in considering an application for a genetically modified food under the European Union's Novel Foods Regulation.

Regulation recognises that providing information for the consumer identifying that GMOs may be present fulfils the labelling obligation.

For foods and food ingredients which are obtained from GMOs, labelling is required when, on the basis of a scientific assessment, it is judged not to be equivalent to an existing food. In determining whether a food is equivalent to a conventional food or not, the regulation draws heavily on the WHO definition of substantial equivalence. From this it is clear that the intention is that the food components compared should be key nutrients and toxicants rather than small fragments of degraded DNA and associated proteins. The regulation also requires labelling if a novel food contains material which is not present in an existing equivalent and which may have implications for the health of some sections of the population. An example would be a novel food that contained a food allergen. Finally, the regulation would require a novel food to be labelled if it contained material which is not present in an existing equivalent foodstuff and which gives rise to ethical concerns. This labelling provision reflects the recommendation of the UK Polkinghorne Committee that foods containing copies of ethically sensitive genes should be labelled.

4.2.3 USA

There are three organisations responsible for controlling genetically modified organisms in the USA. The controls vary depending on the nature of the modification and the consequential properties of the organism.

4.2.3.1 Environmental legislation. The Environment Protection Agency (EPA) has an interest where a plant has been modified to express a pesticidal protein. An example would be maize modified for resistance to the European corn borer. Such crops contain a truncated version of the cry1A(b) gene from *Bacillus thuringiensis* which results in expression of the δ endotoxin CRY1A(b) protein. Such proteins are regarded as pesticides and regulated under the Federal Insecticide Fungicide and Rodenticide Act and the Federal Food Drug and Cosmetic Act. Before such crops can be grown experimentally a permit is required from the EPA. The EPA is also responsible for setting maximum tolerance levels of pesticide residues in foodstuffs. Such tolerances apply equally to levels of a pesticide produced by a GM crop and by externally applied compounds.

In general, a permit is required before any genetically modified crop can be released into the environment or transported between states. The United States Department of Agriculture is responsible for issuing such permits through the Animal and Plant Health Inspection Service (APHIS). The Federal Plant Pest Act (FPPA) authorises APHIS to control interstate movement, importation and field testing of 'organisms and products altered or produced through genetic engineering which are plant pests or which there is a reason to believe are plant pests'. A permit system was introduced in 1987 as an extension of the system which had, for many years, dealt with naturally occurring plant pests. For field tests, an applicant was required to provide a detailed description of the GM crop, details of the proposed experiment and details of measures proposed to ensure that the traits introduced into the crop did not spread into the wider environment. Similarly, a description of the crop and details of containment procedures had to accompany an application for a permit to move a GM crop between states.

Prior to 1992, all field trials were conducted under a permit system. The process of issuing a permit took 120 days from application which included a 30-day state comment period. However, in 1992, the procedures were revised to enable some trials to be conducted under a notification procedure. The change in the rules was designed to ensure that safeguards were proportional to risk and to encourage increased commercial development. Initially, the notification procedure was limited to 6 crops: maize, tomato, soya bean, potato, cotton and tobacco. APHIS considered that there was little or no environmental risk from releases of such crops. For these crops, the revised rules require the provision of information covering details of the crop and experimental sites, the provision of an annual report and the reporting of any adverse effects. APHIS requires notification 10 days before an interstate movement and 30 days before field testing or importation into the US. Up until the end of 1994, experimental releases with maize crops accounted for most notifications. The impact of the notification procedure can be seen in Figure 4.2.

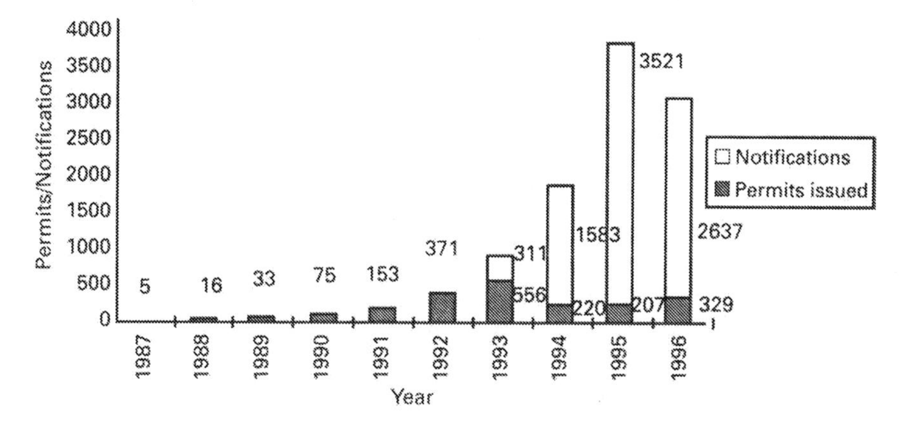

Fig. 4.2 The number of field test sites showing permits issued and notifications acknowledged by APHIS (Animal and Plant Health Inspection Service) in the USA.

4.2.3.2 Food safety. In the USA, the Food and Drug Administration (FDA) has the responsibility for overseeing the safety of food under the Federal Food Drug and Cosmetics Act (FFDCA) which, in a similar way to the UK's Food Safety Act, places the onus on a producer to ensure food is safe. Under the Act, the only foods which require specific approval by the FDA are food additives. The act requires that food additives go through an approval process unless they are considered to be Generally Recognised As Safe (GRAS). GRAS status is demonstrated either by a long history of consumption prior to 1958 or if it is regarded as safe by scientific experts, based on publicly available scientific information. Foods which contain additives that have neither been specifically approved by the FDA nor have GRAS status are considered to be unsafe and their sale is illegal. The FFDCA contains 'adulteration' provisions which authorise the FDA to remove unsafe foods from the market and make producers legally responsible for the safety and wholesomeness of the foods they market.

A policy statement on foods derived from new plant varieties was issued in 1992. In it the FDA concluded that the introduction of new biological techniques remained consistent with the characteristics that underpin the FFDCA (Federal Register, 1992): namely that food safety is determined by the characteristics of the consumed or processed product rather than the method by which it was produced. The FDA also concluded that existing food legislation can adequately address safety issues relating to foods obtained from GM crops. The existing controls impose a clear legal duty on producers to ensure the safety of foods they offer to consumers. This legal duty is backed up by stringent FDA enforcement powers for product recalls and authority to require pre-market-reviews, and approval if necessary, to protect public health.

The statement makes it clear that substances introduced into food by way of plant breeding are considered to be 'added' and thus subject to the food additives provisions of the FFDCA. That is, if they are not GRAS they need to be approved before they can be legally sold. However, since most transgenes and their associated proteins have a history of food use they would generally be regarded as GRAS. The FDA policy statement concluded that a special review of a genetically modified food would only be needed if the food raised safety concerns. In order to clarify when an evaluation may be required, the FDA policy statement identified the following situations as indicators that could trigger an evaluation:

- unexpected effects (produces unexpected genetic effects);
- known toxicants (has significantly higher levels of toxicants than present in other edible varieties of the same species);
- nutrients (significantly altered levels of important nutrients);
- new substances (differs significantly in composition from such substances currently found in food);
- allergenicity (contains proteins that cause an allergic response);
- antibiotic resistance selectable markers (contains marker genes that theoretically may reduce the therapeutic effects of clinically useful antibiotics);
- plants developed to make speciality non-food substances (plants developed to make substances like pharmaceuticals or polymers that will also be used for food);
- issues specific to animal feeds (significant changes in nutrients or toxicants).

Although most GM foods do not require specific approval, the FDA encourages companies to notify details of their products on a voluntary basis. The emphasis remains on the company conducting a safety assessment of the product and communicating the results of the assessment to the FDA. To assist companies assessing the safety of their products, the FDA has developed a series of decision trees. The decision trees focus the evaluation on the differences between the new variety and its unmodified equivalent. This approach is consistent with the concept of substantial equivalence developed by the OECD.

The 1992 policy statement by the FDA indicated that consultations are an appropriate forum for industry and the FDA to address scientific and regulatory issues prior to marketing. Once any necessary consultation has been concluded and safety and regulatory issues have been resolved, the FDA recommend that the product developer should submit to the FDA a summary of the safety and nutritional assessment of the product and make a scientific presentation of the data supporting the safety assessment to

Table 4.2 Genetically modified crops completing the regulatory process in the USA as of March 1997

Crop	Trait	Company
Cotton	Bromoxynil tolerance	Calgene
Cotton	Insect resistance	Monsanto Co
Cotton	Glyphosate tolerance	Monsanto Co
Cotton	Sulphonylurea tolerance	DuPont
Maize	Insect resistance	Ciba Geigy Corp
Maize	Glufosinate tolerance	AgrEvo Inc
Maize	Glufosinate tolerance	Dekalb Genetics Corp
Maize	Male sterile	Plant Genetic Systems
Maize	Insect resistance	Monsanto Co
Maize	Insect resistance, glyphosate tolerance	Monsanto Co
Maize	Insect resistance	Northrup King
Oilseed rape	Glyphosate tolerance	Monsanto Co
Oilseed rape	Glufosinate tolerance	AgrEvo Inc
Oilseed rape	High laurate	Calgene
Oilseed rape	Male sterility/fertility restorer	Plant Genetic Systems
Potato	Insect resistance	Monsanto Co
Soya beans	Glyphosate tolerance	Monsanto Co
Squash	Virus resistance	Asgrow Seed Co
Tomato	Modified ripening	Agritope Inc.
Tomato	Modified ripening	DNA Plant Technology
Tomato	Modified ripening	Monsanto Co
Tomato	Modified ripening	Zeneca Plant Science
Tomato	Delayed ripening	Calgene

FDA scientists. The emphasis placed on the desirability of early pre-market consultations has so far been followed by all companies developing GM foods. Details of GM crops that have completed the regulatory process in the USA are given in Table 4.2.

4.2.3.3 Labelling. The 1992 FDA policy statement also addresses the labelling of foods obtained from new plant varieties. The FDA philosophy is that genetic modification techniques do not result in foods that differ as a class from foods developed through other methods of plant breeding, thus there is no need to label the technique used to produce the food. Provided the company's safety assessment demonstrates that there are no unacceptable differences between the GM variety and the conventional equivalent with respect to toxicants, allergens or nutritional value, no labelling is required. Labelling is required in situations where a food obtained from a GM variety differs from that obtained from a conventional counterpart in such a way that the usual name would be misleading. Labelling would also be required if the modification gave rise to safety issues to which the consumer should be alerted.

4.2.4 Canada

4.2.4.1 Environmental legislation. Canada has been developing legislation relating to the risk assessment of GM crops for a number of years. In December 1992, the Federal Cabinet approved a framework for the regulation of the products of biotechnology in Canada. The principles were intended to ensure that the practical benefits of biotechnology were balanced against the need to protect the environment, human health and safety. As part of this framework, it was agreed to use existing legislation and regulatory institutions. In effect, those government departments that were responsible for regulating products obtained by traditional means would also be responsible for regulating products obtained through the application of biotechnology.

The first set of regulations come into force in January 1995. The regulations provided powers to regulate agricultural products under the Seeds Act, Feeds Act, Fertilizers Act and Health of Animals Act. Amendments to these regulations, published in the Canada Gazette, Part II, on 8 January 1997, are intended to clarify the fact that the standards laid down under the various pieces of product legislation are as high as those provided for under the Canadian Environmental Protection Act.

All GM crops released into the environment under confined conditions have been regulated under guidelines by Agriculture and Agri-Food Canada (AAFC) since 1988. Under the amendment to the Seeds Regulations, confined release experiments will continue to be regulated by AAFC. Under the existing controls, 665 confined releases of GM plants took place in 1996. The amendments to the regulations recognise that a considerable body of knowledge has been developed from previous releases. Based on eight years experience there was no evidence to suggest that the release of plants with particular traits presents an environmental risk. The amendments to the Seeds Regulations will enable researchers to conduct unconfined releases. Indeed since 1995, 16 such releases have already taken place in Canada.

4.2.4.2 Food safety. In many respects, the Canadian policy on the safety of novel foods is closer to that existing in Europe and other countries which apply direct regulatory oversight since Canada has a pre-market notification requirement, as opposed to the USA where such notification is not a requirement.

In Canada, food safety is regulated under the Food and Drugs Act and Regulations, administered by Health Canada. The regulations contain a number of provisions for controlling the sale of food in Canada. Amongst these measures is a provision for a pre-market notification of certain foods intended for sale, specifically infant formulae, irradiated foods and food additives. Under this provision, those intending to place food of these types on the market are required to submit information on the product to the Health Protection Branch of Health Canada. The information required has to be sufficient to enable a determination to be made as to the acceptability of a food prior to sale.

The pre-market notification approach has been applied to the regulation of foods derived from genetically modified organisms along with other novel foods in Canada. In 1995, Health Canada introduced a proposal to define a novel food under the Food and Drug Regulations and to require a pre-market notification of such foods. The proposal defined novel foods as:

- substances that have previously not been used as food in Canada, or result from a process that has not previously been used for food in Canada;
- existing foods that have been modified by genetic manipulation and exhibit one or more characteristics that were previously not identified in that food, or food that results from production by a genetically manipulated organism exhibiting such new characteristics;
- food containing microorganisms that have previously not been used as food or to process food before; and
- food that is substantially modified from the traditional product, or is manufactured by a process that has been substantially modified from the traditional process.

Under the proposal, genetic modification is defined as 'changing the heritable traits of a plant, animal or microorganism by means of intentional manipulation'. This definition is much broader than the European definition which specifically excludes manipulation using mutagenesis techniques. It should be noted that the focus of the Canadian position is on the fact that the food is novel, and not on the method of development. As such, the definition of genetic modification is intentionally broad in order to recognise that novel characteristics might be introduced by either genetic engineering or by more traditional means, but the safety issues remain the same (e.g. herbicide tolerance in crop plants has been accomplished both by genetic engineering and by traditional mutagenesis).

The proposed regulation would specify that an applicant must submit a notification in writing 90 days before selling or advertising for sale. Notification will need to be made to the Health Protection Branch of Health Canada. Within the 90-day period, Health Canada will be required to consider, on the basis of information provided, whether or not the product is acceptable for sale. The information required includes a description of the novel food together with information about the product's development, manufacturing and storage details. Information is also required on the intended use of the product, the history of use of the product in other countries, the estimated level of consumption and the proposed product label. The proposed regulation does not specify that safety data are required in all cases. Instead, the regulation provides for a request from Health Canada to an applicant for the evidence relied upon to establish the safety of the novel food. Where such additional information is required, an applicant would not be permitted to sell the product until Health Canada has agreed to the acceptability of the product.

A request for the evidence relied upon to establish the safety of a novel food permits Health Canada to conduct a safety assessment for the product prior to authorising its sale. Details of the safety assessment criteria adopted by Health Canada have been published in 'Guidelines for the Safety Assessment of Novel Foods, Volume II, Food Directorate, 1994'. The safety assessment criteria incorporate the food safety assessment concept enshrined in the OECD's principle of substantial equivalence. Details of crops for which regulatory decisions have been made in Canada are listed in Table 4.3. Such decisions indicate that in the opinion of Health Canada's Food Safety Directorate, no objection is taken to the use of such products as food in Canada.

4.2.4.3 Labelling. The main legislation covering food labelling in Canada is the Food and Drugs Act, although additional labelling requirements are covered by regulations under the Consumer Packaging and Labelling Act. The labelling of genetically modified foods has been the subject of much

Table 4.3 Canadian Novel Food Decisions as of March 1997

Crop	Trait	Company
Cotton	Bromoxynil tolerance	Calgene
Cotton	Insect resistance	Monsanto Co
Cotton	Glyphosate tolerance	Monsanto Co
Maize	Imidazolinone tolerance	Pioneer Hi-Bred
Maize	Sethoxydim tolerance	BASF Canada
Maize	Insect resistance	Pioneer Hi-Bred
Maize	Insect resistance	Monsanto Canada
Maize	Insect resistance	Ciba Seeds
Maize	Insect resistance	Northrup King
Maize	Glufosinate tolerance	Dekalb Genetics Corp
Oilseed rape	Hybrid breeding system (male sterility)	Plant Genetic Systems
Oilseed rape	Glyphosate tolerance	Monsanto Canada
Oilseed rape	Glufosinate tolerance	AgrEvo Canada
Oilseed rape	Hybrid breeding system (fertility restorer)	Plant Genetic Systems
Oilseed rape	Imidazolinone tolerance	Pioneer Hi-Bred
Oilseed rape	High oleic acid/low linolenic acid	Pioneer Hi-Bred
Oilseed rape	High laurate and myristate	Calgene
Potato	Colorado beetle resistant	Monsanto Canada
Potato	Colorado beetle resistance (Atlantic and Superior varieties)	Monsanto Canada
Soya beans	Glyphosate tolerance	Monsanto Canada
Tomato	Delayed ripening	Calgene
Tomato	Delayed ripening	DNA Plant Technology
Tomato	Reduced pectin degradation	Zeneca Plant Science

debate in Canada, culminating in a proposal issued by Agriculture and Agri-Food Canada in December 1995. In general, food labelling in Canada is seen as serving three main functions:

1. to ensure adequate and accurate information relative to health, safety and economic concerns and to assist consumers in making food choices;
2. to protect consumers and industry from fraudulent or deceptive labelling, packaging and advertising practices; and
3. to promote fair competition and product marketability.

Canada has been closely following developments on labelling of genetically modified foods by the Codex Alimentarius Committee on Food Labelling (CCFL). The discussion document issued by AAFC in December 1995, which was drawn up following a labelling workshop held in 1994 involving a wide range of participants, is intended to form the basis of Canadian government policy on labelling of GM foods. The discussion document sets out five areas where labelling of GM foods needs to be considered. There is general agreement that if a novel food is likely to give rise to health concerns for susceptible groups, then labelling is essential. Similarly, if there is a significant change in the composition of a product then again the need for labelling is generally accepted. There was widespread agreement on the right of consumers to have access to information that enables them to make informed choices when purchasing food products. While labelling was seen as one source of information, it was recognised that other sources such as in-store leaflets might be more appropriate. It was agreed that general labelling to identify a food as genetically modified would provide no new information on the food's safety and may cause people to overlook other important safety information.

In considering the type of information that should appear on the label, the workshop concluded that labels should be understandable, truthful and not misleading. Information on labels should address any health or nutritional implications and identify any compositional changes of concern to the consumer, rather than focus on the process by which the change was obtained. The idea of negative labelling to indicate that a food had not been genetically modified was not ruled out provided such a statement was truthful and not misleading. The workshop also looked at dietary restrictions of various religious groups in considering whether any specific labelling requirements needed to be introduced. The workshop concluded that religious dietary restrictions were already adequately covered by existing food labelling regulations. Finally the workshop recognised that ultimately the labelling of genetically modified foods needs to be harmonised at international level. In this respect, the CCFL (Codex Alimentarius Committee on Food Labelling) is seen as the best forum for achieving international harmonisation. Based on the outcome of the workshop, the discussion document contains five proposals for labelling genetically modified foods:

1. Labels of novel foods or novel food ingredients derived through genetic engineering must identify the presence of:

   - potential health and/or safety risks for individuals or population segments (e.g. the introduction of food allergens); and/or
   - significant compositional or nutritional changes from the traditional food source which have occurred in the product (e.g. a change in the fatty acid profile of vegetable oil).

2. Unless potentially affecting health and/or safety, or significant nutritional or compositional changes have occurred, the fact that the process of genetic engineering has been utilised need not be identified on food labels.
3. The labelling of novel foods or food ingredients derived through genetic engineering should be understandable, truthful and not misleading.

   Voluntary positive labelling that is intended to indicate that a novel food or novel food ingredient has been produced by, or has particular attributes resulting from, genetic engineering is acceptable, on condition that the claim is not misleading or deceptive and the claim itself is factual. Responsible parties should be in a position to support the claim(s) being made. (Violations would be subject to the enforcement measures and penalties provided under the Food and Drugs Act.)

   Voluntary negative labelling that is intended to indicate that a novel food or novel food ingredient has been produced without the use of genetic engineering is acceptable, on condition that the claim is not misleading or deceptive and the claim itself is factual. Responsible parties should be in a position to support the claim(s) being made. (Violations would be subject to the enforcement measures and penalties provided under the Food and Drugs Act.)
4. Dietary restrictions based on religious requirements is outside the current mandate of government and is being adequately addressed within the present regulatory framework of religious groups. A guideline provision is not therefore required.
5. Canada must consider both domestic and international needs in food labelling matters.

The discussion document is intended to form the basis of Canadian policy for labelling GM foods.

4.2.5 Japan

The regulatory system for controlling GMOs in Japan is similar to that operating in Europe. Legislation exists covering both experimental work with GMOs and food safety of GMOs.

4.2.5.1 Environmental legislation. The Japanese Ministry of Education, Science and Culture issued guidelines in 1979 covering all basic research with GMOs in universities. Similar guidelines were issued by the Science and Technology Agency (STA) in 1990 relating to the experimental cultivation of GM crops. Before a GM crop can be grown experimentally, the STA has to give its consent. The STA follows a step-wise approach similar to the European Commission's framework. Small-scale work with GM plants in contained greenhouses is supervised by the STA. When sufficient information has been generated to enable a safety and environmental risk assessment to be conducted, a company can apply for approval to grow a GM crop on a small scale in field conditions. Stringent safeguards are put in place to ensure that there is no adverse effect on the wider environment. By 1994, only 10 deliberate release experiments had been undertaken. These predominantly involved work with rice although some work has also been undertaken with soya beans. In the case of soya, particular emphasis will have been placed on the possibility of outcrossing with related species since there are several species closely related to soya that exist in Japan.

4.2.5.2 Food safety. The earliest food safety legislation relating to foods produced by biotechnology was introduced by the Japanese Ministry of Health and Welfare in 1991. This legislation consisted of basic principles for the safety assessment of foods produced by biotechnology and guidelines covering the manufacture of such foods. These guidelines did not apply to GM foods consumed as such but did cover foods in which an enzyme obtained from a GMO had been used in the production process. Some of the first enzymes to be considered under the guidelines were chymosin together with various amylases and lipases.

In 1995 the Ministry of Health and Welfare issued revised guidelines for the safety assessment of foods and food additives produced by recombinant DNA techniques. The aim of the guidelines is to establish basic requirements for the safety assessment of GM foods in order to ensure their safety. The guidelines apply to both the GM foods consumed as such and to foods where a GMO has been used in the production process. As with food safety legislation in other countries, the guidelines are based upon the concept of substantial equivalence. The Japanese safety assessment is based upon four categories of information:

1. information on inserted genetic material;
2. information on broad human consumption history;
3. information on components of foods;
4. information on differences in usage between the conventional variety and the new varieties.

A company wishing to manufacture or import a GM food or food additive must first obtain confirmation from the Ministry of Health and Welfare

Table 4.4 Genetically modified crops approved in Japan as of March 1997

Crop	Trait	Company
Maize	Insect resistance	Ciba Geigy Japan
Maize	Insect resistance	Northrup King Co
Oilseed rape	Herbicide tolerance	AgrEvo KK
Oilseed rape	Herbicide tolerance	Monsanto Japan
Oilseed rape	Herbicide tolerance	Plant Genetic Systems
Potato	Insect resistance	Monsanto Japan
Soya beans	Herbicide tolerance	Monsanto Japan

that their product complies with the guidelines. Details of GM crops that have been approved in Japan are given in Table 4.4.

4.2.5.3 Labelling. There has been considerable debate in Japan on the need, or otherwise, for labelling of GM foods. When the most recent guidelines on the safety assessment of GM foods were being drawn up, there was considerable pressure from consumer groups for labelling provisions to be included in the guidelines. However, the guidelines do not contain any reference to labelling requirements. Indeed, of the products that have been approved so far, there has been no requirement for labelling to indicate that the foods or food ingredients have been obtained from a GMO.

4.2.6 Australia and New Zealand

4.2.6.1 Environmental legislation. At present, the research, development and use of novel genetic modification techniques is controlled in Australia by the Genetic Manipulation Advisory Committee. In New Zealand, these functions are controlled by the Interim Assessment Group although this is about to be replaced by the statutory Environmental Risk Management Authority.

4.2.6.2 Food safety. In Australia and New Zealand, food safety issues are covered by the Australia New Zealand Food Authority (ANZFA) under powers contained in the Australia New Zealand Food Authority Act 1991. At present, there are no statutory pre-market clearance procedures for novel foods in Australia and New Zealand. Under voluntary arrangements novel foods can be referred to the ANZFA for assessment. The current voluntary procedures are seen as lacking clear guidance to industry as to whether a novel food is acceptable or not. Equally the current situation is seen as leading to a loss in consumer confidence. In a discussion paper published in 1996, ANZFA proposed either a pre-market approval process or a pre-market notification process (ANZFA, 1996). The pre-market notification process was seen as the most flexible option, with a minimal information requirement to enable an initial assessment. However, if considered necessary, additional information could be requested from a company.

In February 1997, ANZFA issued a proposal draft standard (A18) for food derived from gene technology (ANZFA, 1997a). This was accompanied by an information paper containing assessment guidelines (ANZFA, 1997b) for foods and food ingredients to be included in standard A18. The draft standard does not apply to products such as food additives that are already regulated under the food standards code. The draft standard defines food derived from gene technology as:

1. food which has been modified by gene technology;
2. food containing, or derived from, ingredients or components modified by gene technology.

The term 'gene technoology' is equivalent to the definition of genetic modification contained in the European Commission's Deliberate Release Directive 90/220/EEC. Under the draft standard the sale of food derived from gene technology would be prohibited unless it has been approved following a full safety assessment. The proposed ANZFA approach to assessing the safety of genetically modified foods is based on the concept of substantial equivalence and follows a decision tree similar to the approach developed in the UK. The draft assessment guidelines identify five categories of food:

1. foods consisting of chemically defined substances;
2. less well-defined substances such as fats and oils;
3. foods produced using GMOs where the GMO has been removed from the final product;
4. transgenic plants or animals which contain new or altered genetic material;
5. foods such as yoghurt where the genetically modified fermentation microorganism remains in the food.

4.2.6.3 Labelling. The draft standard contains provisions for labelling genetically modified foods. At the same time, the assessment guidelines identify the important role that other sources of information can play in helping consumers make informed choices. The ANZFA labelling guidelines draw upon the labelling provisions contained in the EC Novel Foods Regulation. However, in addition to the four criteria in the EC regulation, the ANZFA guidelines also require GM foods to be labelled when:

> Any ingredient (containing genetic material from a genetically modified organism) comprises more than five percent of the food at the time of manufacture of the food.

The above criterion would require labelling of foods for which it can be shown that there are no significant differences when compared with a conventional equivalent.

4.2.7 China

In 1993, the state Science and Technology Commission introduced the first piece of legislation covering work with GMOs in contained facilities. The Chinese legislation is based on four categories of risk ranging from non-harmful to highly dangerous. Work with GMOs that falls into the two highest risk categories requires specific approval before the work can commence. Controls on the commercial use of GMOs are not very extensive. Those involved in developing a commercial scale product are only required to conduct an internal risk assessment. For GM crops there is also a requirement to demonstrate that the variety has a value for cultivation in the same way that new plant varieties in Europe have to be tested to ensure that they are distinct, uniform and stable.

In the last 10 years, China has placed considerable emphasis on genetic modification to introduce insect resistance, disease resistance and improved nutritional quality. Small-scale field trials have been conducted on GM potatoes. The only GM crops that are known to be grown commercially in China at present are a herbicide-tolerant tobacco and a GM soya bean variety.

4.3 Moves towards international harmonisation

4.3.1 Environmental legislation

At the 1992 United Nations Conference on Environment and Development held in Rio de Janeiro, Agenda 21 established a comprehensive programme of work to ensure sustainable development in the 21st century. Amongst the planned activities was a commitment to the environmentally sound management of biotechnology through the development of international safety guidelines. Working in conjunction with The Netherlands, the UK developed a set of draft International Technical Guidelines on safety in biotechnology. These draft guidelines were then taken forward by the United Nations Environment Programme (UNEP) together with a programme to assist countries to develop and strengthen their capacities to achieve safety in biotechnology. Following seven regional consultation meetings, the guidelines were adopted at a global consultation meeting held in Cairo in December 1995.

In addition to the development of guidelines, Agenda 21 called on Governments to consider the need for a biosafety protocol to the Convention on

Biological Diversity. As part of the process of developing such a protocol, an open-ended ad-hoc working group was established by the Conference of the Parties to the Convention on Biological Diversity. The first meeting of the ad-hoc working group was held in Aarhus, Denmark in 1996. The Biosafety protocol is intended to focus on the transboundary movement of any living modified organism (LMO) resulting from modern biotechnology that may have an adverse effect on the conservation and sustainable use of biological diversity. In drawing up the protocol, consideration will be given to the development of appropriate procedures for advanced informed agreement before the transboundary movement of LMOs. The Biosafety Protocol is intended to enable the global community to derive maximum benefit from the potential of biotechnology to contribute to sustainable development while ensuring that biotechnology is developed and applied within a sound international framework to ensure safety. The intention is that a draft of the protocol will be ready for consideration in mid 1998.

4.3.2 Food safety

The first steps towards international harmonisation of the food safety assessment of GM foods were taken by the FAO and WHO in 1990. They convened an expert consultation on the 'Assessment of Biotechnology in Food Production and Processing as Related to Food Safety'. A number of recommendations flowed from the 1990 consultation (WHO, 1991). It was recommended that the safety assessment of foods produced by biotechnology should take into account the molecular, biological and chemical characteristics of the food under assessment. It was recognised that these characteristics will determine the need for and extent of any toxicological assessment. The consultation recognised the limitations of traditional toxicological test methods when applied to whole foods and recommended that a more structured approach to safety assessment should be developed. One of the key conclusions was that the use of genetic modification techniques 'does not result in food which is inherently less safe than that produced by conventional means'.

The 1990 consultation identified the comparative principle, whereby the food being assessed is compared with one that has an accepted level of safety, as being of considerable importance. In 1991, the OECD expanded upon this comparative principle and formulated the concept of substantial equivalence (OECD, 1993). The concept codifies the idea that if a food or food ingredient under consideration can be shown to be essentially equivalent in composition to an existing food or food ingredient then it can be assumed that the new food is as safe as the conventional equivalent. The WHO and FAO refined the concept at an expert consultation meeting held in Rome in 1996. In the report of this meeting, substantial equivalence was identified as being 'established by a demonstration that the characteristics assessed for the genetically modified organism, or the specific food product derived therefrom, are equivalent to the same characteristics of the conventional comparator. The levels and variation for characteristics in the genetically modified organism must be within the natural range of variation for those characteristics considered in the comparator and be based upon an appropriate analysis of data' (FAO/WHO, 1996).

Before a comparison can be undertaken, it is necessary to characterise the GM variety to ensure that the appropriate characteristics are assessed. The 1996 FAO/WHO report identifies a number of pieces of information that will be of use in this respect. In addition to details of the host organism and details of how the host has been modified it is necessary to characterise the food product itself. It is essential to look not only for intentional changes but also to consider any unintentional changes. In characterising the food product, it is important to consider both phenotypic characteristics and compositional analysis. The type of phenotypic characteristics assessed for a GM plant would include crop morphology, growth, yield and disease resistance. In assessing the composition of the GM product, the FAO/WHO report identified the need to consider key nutrients and toxicants of the food in question. The report also commented that 'analysing a broader spectrum of components is generally unnecessary, but should be considered if there is an indication from other traits that there may be unintended effects of the genetic modification'.

Following a comparison of the GM product with a conventional counterpart, three outcomes are possible:

1. the GMO or food product obtained from it is substantially equivalent to a conventional counterpart;
2. the GMO or food product obtained from it is substantially equivalent to a conventional counterpart except for a few clearly defined differences;
3. the GMO or food product obtained from it is not substantially equivalent to a conventional counterpart – either because the differences cannot be defined or because there is no existing counterpart to compare it with.

Where a food can be shown to be substantially equivalent, it is considered to be as safe as its counterpart and no further safety assessment is required. Where there are clearly defined differences between the GM food and its conventional counterpart, the safety implications of the differences need to be fully assessed. Where a food is not substantially equivalent it does not mean that the food is unsafe. However, there would be a need for extensive data to be provided before the product's safety can be fully assessed.

Many countries have already incorporated the concept of substantial equivalence into their regulatory framework. However, in an attempt to facilitate greater harmonisation between regulatory approaches, the OECD is holding additional workshops, a recent example being the workshop on the toxicological and nutritional assessment of novel foods which took place at Aussois, France, in March 1997.

4.3.3 Labelling

With so many foods traded internationally, there is a clear need to reach a harmonised approach to labelling GM foods if potential trade barriers are to be eliminated. The Codex Committee on Food Labelling (CCFL) is addressing the implications of biotechnology for food labelling. At the 1994 Codex meeting in Ottawa, a discussion paper prepared by the USA was considered (FAO/WHO, 1996a). Discussions identified two main approaches to labelling. A number of countries advocated comprehensive labelling of all products produced by modern biotechnology. They suggested that failure to provide full labelling would deny consumers the right to choose whether to buy such products or not. Equally, many representatives were of the view that labelling should be used to identify safety concerns, including allergenicity, provide nutritional information and to identify the composition of the food product. Recognising the divergent views on labelling, the CCFL decided to seek the advice of the Codex Executive Committee on how labelling guidelines should be formulated, especially in view of the four Statements of Principle of the Role of Science in the Codex decision making process.

At the 42nd session of the Codex Executive Committee it was agreed that the Codex Commission's four Statements of Principle concerning the Role of Science should be closely adhered to (FAO/WHO, 1997b). These four principles are:

1. Food standards, guidelines and other Codex recommendations should be based on sound scientific analysis and evidence, involving a thorough review of all relevant information, in order that the standards assure the quality and safety of the food supply.
2. When elaborating and deciding upon food standards Codex will have regard, where appropriate, to other legitimate factors relevant for the health protection of consumers and for the promotion of fair trade practices in food trade.
3. In this regard it is noted that food labelling plays an important role in furthering both of these objectives.
4. When a situation arises that members of Codex agree on the necessary level of protection of public health but hold differing views about other considerations, members may abstain from acceptance of the relevant standard without necessarily preventing the decision by Codex.

The Executive Committee recognised that some consumers claim the right to know whether food has been produced by biotechnology. However, it was considered that this right was ill-defined and could not be used by

Codex as the primary basis for reaching decisions on appropriate labelling. The Committee identified the main factors to be considered when reaching decisions on labelling as the need to protect consumer health, nutritional changes, significant technological changes in the properties of the food itself and the need to prevent deceptive trade practices.

The paper considered at the 1997 meeting of the CCFL proposed that for the purposes of labelling, the definition of a GMO should be that defined in the EC Directive 90/220/EEC (FAO/WHO, 1997). The paper also recognised that the Codex food labelling mandate relates to the food itself, its safety, characteristics, nutritional composition or intended use. In the case of GM foods, it is the intention that the consumer is provided with clear information for any new product obtained through biotechnology which has specific characteristics not found in conventional foods. It was proposed that the Codex General Standard for the Labelling of Pre-packaged Foods should be revised to include a section on labelling of foods obtained through biotechnology in the mandatory requirements section. The proposal put to the 1997 CCFL meeting would require labelling of products obtained through biotechnology, which were defined as foods composed of or containing GMOs, in the following situations:

1. When a food or food ingredient obtained through biotechnology is no longer substantially equivalent to the corresponding existing food or food ingredient as regards:

   - composition
   - nutritional value
   - intended use

The characteristics which make it different from the reference food should be clearly identified in the labelling. In particular, if the nutrient content is significantly modified, relevant nutrient declaration should be provided in conformity with the Guidelines for Nutrition Labelling. If the mode of preparation is significantly different from that for the equivalent food, clear instructions for use should be provided.

2. When a food produced by biotechnology is not substantially equivalent to any existing food in the food supply and no conventional comparator exists.

The labelling shall indicate clearly the nature of the product, its nutritional composition, its intended use, the method by which it was obtained, and any other essential characteristic necessary to provide a clear description of the product.

3. The presence in a food obtained through biotechnology of material from the sources which may pose religious or ethical concerns and which is not present in an existing equivalent foodstuff shall always be declared.

4.4 Conclusions

The application of genetic modification techniques to food production is a relatively recent development, although biotechnology has been used in food production for many centuries and indeed genetic modification has formed the basis of selective plant and animal breeding over the same time period. The main change is that today the results of such selective breeding can be realised on a much shorter time scale. It is also possible to be more selective as to which genes are transferred and to transfer genes between species. In response to concerns about the lack of familiarity with the behaviour of GMOs, many countries now have procedures in place to ensure the safety of GM foods. Although each country has developed its own approach, there are many elements in common.

Controls in place to ensure that GMOs do not harm the environment are similar in all the countries considered in this chapter. Experience gained from small-scale trials is taken into account when considering applications for larger-scale trials. In some countries, the regulatory authorities have concluded that they now have sufficient information on the behaviour of some crops to enable a notification rather than approval procedure to be adopted.

There is much in common between the approaches adopted by different countries to the assessment of the food safety of GM crops. Many countries have adopted a decision tree approach based on the decision tree originally developed in the United Kingdom. This decision tree is used to identify the information required to support an application for food safety clearance. All the countries considered in this chapter use the concept of substantial equivalence in assessing the safety of GM foods.

In the case of labelling, there is, again, much in common between different countries. There is general recognition of the need to label foods where the food may give rise to safety concerns such as allergenicity. The right of consumers to information which will enable them to make informed choices is also acknowledged. However, there is an active debate on the most appropriate means of providing such information. In considering the provision of information on product labels, account needs to be taken of the practicalities of labelling.

There are already a number of initiatives underway to develop internationally harmonised approaches to controlling GMOs. These will become ever more important as the range and number of GM foods reaching commercial development increases in decades to come.

References

Advisory Committee on Novel Foods and Processes (ACNFP) (1994) Annual report. Ministry of Agriculture, Fisheries and Food Publications, London SE99 7TP, UK.

Agriculture and Agri-Food Canada (1995) Labelling of novel foods derived through genetic engineering. Food Inspection Directorate, 1 December 1995.
Australia New Zealand Food Authority (1996) The safety assessment of novel foods and novel food ingredients, Discussion and options paper, September 1996.
Australia New Zealand Food Authority (1997a) Draft variation to the food standards code, Standard A18 – food derived from gene technology, 5 February 1997.
Australia New Zealand Food Authority (1997b) Assessment guidelines for foods and food ingredients to be included in standard A18 – food derived from gene technology. Information paper. February 1997.
Council Directive of 23 April 1990 on the deliberate release into the environment of genetically modified organisms 90/220/EEC. Official Journal of the European Communities No. L117/15, 8 May 1990.
Council Directive on the Contained Use of Genetically Modified Microorganisms 90/219/EEC (1990) Official Journal of the European Communities No. L117/1, 8 May 1990.
Department of Health (1991) Report on Health and Social Subjects 38 Guidelines on the Assessment of Novel Foods and Processes, Advisory Committee on Novel Foods and Processes, HMSO, London, UK.
FAO/WHO (1994) Food Standards Programme, Codex Committee on Food Labelling, Implications of Biotechnology for Food Labelling, CX/FL/94/8.
FAO/WHO (1996) Biotechnology and Food Safety, Report of a joint FAO/WHO consultation Rome 30 September–4 October 1996, FAO Food and Nutrition Paper 61.
FAO/WHO (1997a) Food Standards Programme, Codex Alimentarius Commission, Report of the twenty fourth session of the Codex Committee on food labelling, Alinorm 97/22.
FAO/WHO (1997b) Food Standards Programme, Codex Committee on Food Labelling, Recommendations for the Labelling of Food Obtained through Biotechnology. CX/FL/97/7.
Federal Register Vol. 57 No. 104 Friday 29 May 1992, Food and Drug Administration. Statement of Policy: Foods Derived from New Plant varieties.
Health Canada (1995) Proposed amendment to the Food and Drug Regulations (Schedule No. 948), Canada Gazette Part I, 26 August 1995.
Japanese Ministry of Health and Welfare (1995) Revision of guidelines for safety assessment of foods and food additives produced by the recombinant DNA techniques. October 1995.
Ministry of Agriculture, Fisheries and Food (1993) Report of the Committee on the Ethics of Genetic Modification and Food Use, HMSO, London, UK.
OECD (1993) Safety Evaluation of Foods Produced by Modern Biotechnology – concepts and principles, OECD, Paris.
Regulation (EC) No. 258/97 of the European Parliament and of the Council of 25 January 1997 concerning Novel Foods And Novel Food Ingredients, Official Journal of the European Communities No L 43, 14 February 1997.
WHO (1991) Strategies for assessing the Safety of Foods Produced by Biotechnology. Report of a joint FAO/WHO consultation, WHO Geneva.

5 Communicating biotechnology to an uncertain public: the need to raise awareness

JOANNA SCOTT

5.1 Introduction

In today's technological age, some may wonder why the introduction of genetically modified foods should be an issue. We take flying in an aircraft for granted. We do not worry about the use of lasers in the home. So why should we worry about a new method of food production?

Historically, new technologies have always been treated with some suspicion to a greater or lesser extent; be it fear of the unknown or concerns about the possible consequences such as job losses or damage to the environment. Furthermore, when a problem occurs today, live footage of the event is relayed around the world in minutes. Access to information and the power of the media ensures that everyone has an opinion, whether based on well-informed choice or not.

Consumers are increasingly reluctant to accept the risks of new technology unless direct consumer benefits are immediately apparent. Consumer expectations are high, and, if these are not met, the product or service on offer is quickly rejected. Meeting consumer demand for information, especially if related to a complex technical issue such as food biotechnology, represents a major challenge.

Recent history has shown that the food industry and its products are subject to intense scrutiny by the media and consumer groups. Food scares and heightened awareness of health and nutrition have put food high on the agenda of subjects of intense consumer interest. As the products of gene technology reach the marketplace, media coverage and consumer surveys confirm that modern biotechnology is in the process of becoming a major consumer issue.

It is beyond the scope of this chapter to review the communication initiatives of all bodies – such as government departments, academic institutions, educators and consumer groups – with an interest in biotechnology. Nevertheless, industry should examine the activities of such groups, on a country by country basis, whenever a biotechnology communications programme is being planned. This is important so that opposing views can be taken into account and to avoid duplication of effort. Where complementary activities are being considered, the benefits of working in partnership should not be overlooked. Furthermore, the psychological mechanisms underlying consumer acceptance should also be taken into account whenever possible (these are discussed in more detail in Chapter 2).

5.2 Attitudes to biotechnology

In the UK, the Food and Drink Federation (FDF), an industrial trade association, has recognized the need to respond to consumer questions and concerns about biotechnology. As a first step towards developing better communication channels with consumers, the FDF commissioned a National Opinion Poll in 1995 to establish the level of knowledge and attitudes about biotechnology in Great Britain. The poll consisted of face-to-face interviews with 1700 adults (FDF, 1995).

Respondents in the FDF survey were initially asked a very general question about how much they knew about biotechnology. The results showed that 68% of respondents admitted to not knowing anything about biotechnology or never to have heard of the term. Fewer than 1% claimed to know 'a great deal', 6% said they knew 'a fair amount' and just over a quarter said they knew 'not very much'. Those who had claimed any knowledge of biotechnology were then asked, without prompting, to name areas where they thought biotechnology was currently applied most. In response, medical and plant applications were mentioned most often but nearly one third were unable to provide any answer at all. The group claiming to know 'a fair amount' about biotechnology was then presented with a list of specific food products and processes and was asked to identify those involving biotechnology. Nearly a quarter of the respondents in this group incorrectly identified the canning of peas as an application of biotechnology. The survey clearly indicated that even those consumers who considered themselves 'better informed' about biotechnology did not know very much about it.

Ethical concerns and potential safety issues were identified as the principal disadvantages of the new biotechnology by 18% and 17% of the respondents, respectively. In general, respondents were more likely to support developments in plant as opposed to animal applications. Less than one-third of all respondents were fairly or extremely confident that the regulatory controls of modern biotechnology were adequate. Importantly, 93% of respondents thought that products which are the result of modern biotechnology should be clearly labelled. When asked whether or not they were likely to buy the products of modern biotechnology, 23% of all respondents said that they would avoid them but a large majority (67%) indicated an interest in buying such foods. However, a large proportion of consumers (45%) also indicated a need for more information prior to purchase.

The FDF survey also examined consumers' attitudes towards information sources about biotechnology. Of those who indicated that they had received such information (two-thirds of adults had received no information at all), 17% received information from TV and 15% from newspapers and magazines. Notably, the group claiming to know 'a fair amount' about biotechnology also claimed to have received information from a wider range of sources such as scientific magazines and educational establishments. At the time of the survey (1995), none of the respondents had received information from consumer groups, the Government or from supermarkets or other retailers.

Respondents in the FDF survey were also shown a list of groups and organizations and asked which two or three should be doing the most to provide information on modern biotechnology. The results (shown in Table 5.1) indicated that the majority of consumers felt that the Government and the food industry should supply information. Yet, when asked which of the same groups would provide the most trustworthy information, the Government and the food industry were not perceived as very trustworthy; health professionals and consumer groups were seen as more trustworthy.

The FDF survey in the UK was in broad agreement with many other consumer surveys carried out in the USA, Europe and Japan in the early 1990s (Eurobarometer, 1993; Hamstra, 1993; Hoban, 1996; Hoban and Kendall, 1992; IGD 1995 and 1995a; Marlier, 1992). For a more detailed discussion of cross-cultural and individual differences between consumers, the reader is referred to Chapter 2. In all these surveys, consumers have expressed quite clearly their requirement for more information regarding biotechnology. The FDF therefore concluded that there was a real need to raise consumer awareness about modern biotechnology. Furthermore, if consumer needs for information and reassurance were properly met, many of

Table 5.1 Organizations who should provide information about biotechnology and their trustworthiness as assessed by a survey of 1700 adult consumers in the UK. Adapted from Food and Drink Federation (1995), with permission

Organization	'Should provide advice' (% of consumers agreeing with this statement)	'Are trustworthy' (% of consumers agreeing with this statement)
Government	46	10
Food manufacturers	31	3
Media	31	6
Retailers	28	3
Scientists	21	11
Consumer groups	19	14
Health professionals	18	18
Healthcare companies	15	5
Environmental groups	13	10
Schools	11	4

the non-contentious products of gene technology (i.e. those raising no ethical concerns) could achieve consumer acceptance. Finally, the FDF also concluded that working in partnership with more trusted organizations would be the most effective way forward for a consumer education programme.

5.3 The FDF communication programme on biotechnology

The FDF's next task was to develop core materials to assist the communication process. A 30-page colour booklet – *Food for Our Future* – was produced to explain the technology and its applications in food and to address key concerns. This booklet was not intended to answer all questions; instead, it was designed to offer a starting point for open and constructive debate. The booklet was aimed at the informed consumer. During production of the booklet, several key pointers to successful communication emerged:

(a) It was important to let the reader know from what standpoint the booklet was written, i.e. that this was a food industry publication.
(b) It was important to acknowledge alternative views and address specific concerns about biotechnology. Thus, one-third of the booklet was devoted to consumer concerns such as safety, the environment, ownership of rights, benefits for developing countries, labelling, ethical issues, and animal welfare.
(c) The text needed to be easily understood and was broken up into chunks with bullet points and headings.
(d) The language of the text was carefully chosen. For example, the term 'genetic modification' was deemed more acceptable than 'genetic engineering' or 'manipulation'. Consistency of language was also important.
(e) Visual material was crucial for adding interest and breaking up the text; it, too, had to be appropriate for the target audience.

The booklet was included in the resource packs at several conferences and exhibitions held in the UK in 1996 and was distributed within the industry and to other interest groups on demand. Feedback to date has underlined the importance of addressing all concerns with clarity and openness and recognizing that the acceptability of novel food products to the consumer is a matter of individual choice. The alternative approach of describing the technology as an extension of natural processes and avoiding direct reference to gene technology has been shown to arouse suspicions in the consumer and is consequently seen as a less desirable method of achieving consumer acceptance (IGD, 1995). Similarly, in the USA, the American

Dietetic Association has advised that 'Biotechnology needs only to be explained, not promoted' (ADA, 1993).

The FDF launched a communication initiative entitled FoodFuture in the summer of 1995. The aim was to promote the understanding of modern biotechnology, and in particular of genetic modification, through frank and honest dialogue with all sectors of society. These sectors included educationalists, religious groups, representatives from farming, manufacturing, retailing and regulatory bodies, research organizations, consumer and women's groups, and members of the medical and paramedical professions. Thereafter, direct communication with consumers would be planned.

The core feature of the FoodFuture programme was a series of forum events which aimed to generate widespread discussion about the key applications and issues relating to genetic modification. The first three events took place around the UK in York, Reading and Nottingham. In each case the forum event was co-hosted with other organizations: the York forum with Nestlé and York Bioscience; the Reading forum with the National Association of British and Irish Millers and the Government-sponsored Institute of Food Research; and the Nottingham forum with the Biotechnology and Biological Sciences Research Council (a Government research funding body) and the University of Nottingham. The programme for each one-day event was the same although different speakers covered the same topics at different events. One of the presentations at each event was devoted to consumer perspectives and was given by a key representative of a consumer organization. Other topics covered in each forum included research, regulatory aspects, farming, manufacturing, retailing and ethics. The audiences included local consumer, women's, youth, environmental and religious groups as well as educators and students. Members of the UK Government's Food Advisory Committee, the Ministry of Agriculture, Fisheries and Food, and the Advisory Committee on Novel Foods and Processes also attended. (For a more detailed discussion of these bodies and their role in the regulation of biotechnology, the reader is referred to Chapter 4.)

The key points which recurred at these events were the need to:

1. ensure that the views of minority groups are considered;
2. reassure the consumer that regulations are adequate, that products are safe and that, if released, they will not damage the environment;
3. inform the consumer that foods are the result of modern biotechnology thereby allowing the exercise of choice.

The FDF events described above culminated in the half-day National Review Conference on 'Food and Modern Biotechnology' held in London in March 1996. The invited audience reflected the range of companies and organizations with an interest in the future development of modern biotechnology. The three key papers drew together the issues raised at previous forum events and covered the technical challenges facing food biotechnology, the ethical issues and information and labelling considerations. Again, the event provided an opportunity for a wide-ranging debate.

Since then, FDF has produced further written materials and an Internet Web Site to support the introduction of new genetically modified foods onto the market and, in conjunction with the Science Museum, is developing a UK touring exhibition to explain biotechnology to a broader audience. A main focus has been working with other industry members, government and the media to ensure wide availability of factually accurate information. In particular FDF has played an active role in the debate on labelling. Future information initiatives are also being planned for both consumers and schools.

5.4 Other industry initiatives in the UK

The Institute of Grocery Distribution (IGD) is also involved in the communications process on behalf of UK food retailers. The IGD has held a number of consumer focus groups and carried out a consumer survey of public perceptions of biotechnology (IGD, 1995a). This has been followed by a number of training workshops for industry members including one on Communication Solutions in Marketing the Food Products of Gene Technology. In addition, the IGD has produced a glossary of terms – a document proposing the use of a consistent set of terms and definitions by communicators and a Factfile for non-technical industry managers (IGD, 1995b, 1996/1997).

The launch of the first products of modern food biotechnology in the UK inevitably drew the retailers into the communication process. In January 1994, a vegetarian cheese produced using the enzyme chymosin from a genetically modified microorganism was introduced into the UK market by Co-op. Initially, it was introduced unlabelled but, subsequently, labelling was added to justify the vegetarian claim. Shelf cards and explanatory leaflets were also made available. In February 1996, Sainsbury's and Safeway's introduced Zeneca's tomato purée in cans clearly labelled as made from genetically modified tomatoes. Shelf cards and consumer leaflets were also provided, as well as consumer 'hotlines' for additional information if required. Similarly, during 1997, consumer information leaflets were provided in response to the introduction of genetically modified soya.

5.5 The issue of labelling

A recurrent theme in all consumer surveys to date is the consumer's wish for all products of gene technology to be so labelled thereby facilitating informed choice (European Commission, 1995; Consumers International, 1995; etc.). While manufacturers are generally willing to supply consumers with the information they want, there are considerable practical difficulties in labelling every single food whose production process might have involved genetic modification. For example, should a food product (e.g. a pizza) containing the genetically modified tomato paste as an ingredient, even if present at very low levels, be labelled? The issue becomes even more complicated when commodity products from a wide range of suppliers are bulked (as is the case for soybeans). Furthermore, to the extent that labelling of genetically modified foods is legally required, enforcement could be a major technical issue, especially for products in which all the genetic material has been removed or destroyed by processing or cooking.

Whatever the outcome of the current debate on labelling (for more details on this issue, the reader is referred to Chapters 2 and 4), it has been recognized that a label cannot provide all the facts. There remains the need to communicate the wider aspects of biotechnology to a range of audiences.

5.6 International communication initiatives

In Switzerland, a biotechnology communications programme was established by a partnership of industry, Government, consumer and environmental groups. Together, they developed a touring exhibition entitled 'Gentechnik: Pro und Contra' addressing questions such as: What is genetics? What is gene technology? How is gene technology applied? What are the issues/concerns/areas of agreement? At the end of the exhibition an information desk provided a selection of leaflets produced by the partners. The exhibition was targeted at schools, local authorities, Government, industry and the media. In parallel with the exhibition, open public fora took place when two speakers with opposing perspectives were invited to debate their views. This was followed by a facilitated discussion. The exhibition began touring Switzerland in 1993 and in the first two years, visited six towns and was seen by nearly 200 000 people. It was usually sited in science centres, museums and shopping centres. Additional communication initiatives in other European countries are also discussed in Chapter 2.

In the USA, several industry initiatives for communicating biotechnology are in place. For example, the International Food Information Council (IFIC) based in Washington DC produced a 12-page booklet entitled *Food Biotechnology – Health and Harvest for our Times* (IFIC, 1995). The booklet explains both traditional and modern biotechnology, and covers labelling and safety issues. In addition, companies such as Calgene have launched extensive product-specific promotional campaigns to educate consumers about genetically modified foods. For example, the FlavrSavr® tomato with improved flavour was introduced into supermarkets amidst a blaze of publicity and accompanied by consumer leaflets explaining the benefits of the product and the reasons for its higher price compared with the traditional tomato. Despite opposition by some consumer groups, the new tomato was generally well-received by consumers.

5.7 Conclusions

Consumers currently have little knowledge about biotechnology but have a clear desire to be well-informed about the products of biotechnology. In particular, labelling of genetically modified foods is demanded.

The benefits of biotechnology need to be clearly communicated to the consumer. There must be an open and honest debate with the risks and downsides being made clear. It is not enough to tell consumers that a product is safe and expect them to buy it; they have to be persuaded to choose to buy it.

Consumer acceptance of the products of modern biotechnology may differ according to the product and from country to country. Reactions to the first products will influence reactions to those that follow, for better or for worse.

Consumer information programmes should take account of the timing of proposed product launches, be accurate and honest, be relevant to the target audience, contain the right level of detail and be communicated through appropriate media. Product information leaflets should explain and reassure by giving the full story: what the product is, what it offers, why it is safe, how to identify the foods, together with an invitation to find out more. Clearly, leaflets need to be written in a language that is comprehensible to the consumer. At all stages of the communications process, survey evidence is important to check understanding and acceptance.

Working in partnership with more trusted organizations is the most effective way forward. In particular, it is essential for scientists and other experts to be part of the communication process – especially scientists who can communicate clearly and effectively.

The popular media will ultimately be one of the more important conduits of information to the consumer. Provision of information is not the end of the process, it is only the beginning. It is the starting point of a dialogue with the consumer, in which their concerns are listened to and responded to.

It is important to be aware of the limitations inherent in any information campaign. Where consumer concerns are based on incomplete or inaccurate information, providing the right information can change their views. However, where consumer concerns are based on intrinsic beliefs, such as religious issues, no amount of information will make any difference to that consumer's attitude.

References

American Dietetic Association (1993) Position of the ADA: Biotechnology and the future of food. *Journal of the American Dietetic Association*, **33** (2), 189–91.
Consumers International (1995) *Food of the Future: The Risks and Realities of Biotechnology*. Conference Proceedings. Consumers International, London, UK.
Department of Trade and Industry (1996) *Crusade for Biotechnology Prospectus 96–7*, DTI, London, UK.
Eurobarometer (1993) *Biotechnology and Genetic Engineering: What Europeans think about it in 1993*. Commission of the European Communities, Brussels, Belgium.
European Commission Group of Advisors on the Ethical Implications of Biotechnology. (1995) *Ethical Aspect of the Labelling of Foods derived from Modern Biotechnology*. European Commission, Brussels, Belgium.
Food and Drink Federation/National Opinion Poll (1995) *Consumers and Biotechnology: Report on Research Conducted for the FDF*, FDF, London, UK.
Food and Drink Federation (1995) *Food for Our Future*. FDF, London, UK.
Hamstra, A. M. (1993) *Consumer Acceptance of Food Biotechnology: The Relation Between Product Evaluation and Acceptance*. SWOKA Research Report, The Hague, The Netherlands.
Hamstra, A. M. (1995) *Consumer Acceptance Survey*. SWOKA Research Report, The Hague, The Netherlands.
Hoban, T. J. (1996) How Japanese consumers view biotechnology. *Food Technology*, July, 85–8.
Hoban, T. J. and Kendall, P. A. (1992) *Consumer Attitudes about the Use of Biotechnology in Agriculture and Food Production*. North Carolina State University, Raleigh, NC, USA.
Institute of Grocery Distribution (1995) *Qualitative Consumer Research*. IGD, Watford, UK.
Institute of Grocery Distribution (1995a) *Consumer Attitudes to Biotechnology in Agriculture and Food: A critical Review*. IGD, Watford, UK.
Institute of Grocery Distribution (1995b) *Biotechnology Glossary of Terms*. IGD, Watford, UK.
Institute of Grocery Distribution (1996/1997) Biotechnology Factfile. IGD, Watford, UK.
International Food Information Council (1995) *Food Biotechnology – Health and Harvest for our Times*. IFIC, Washington DC, USA.
Marlier, E. (1992) Eurobarometer 35.1: Opinions of Europeans on biotechnology in 1991. In: Durant, J. (ed.) Biotechnology in Public: A Review of Recent Research. The Science Museum, London, UK.

Part 2
Case Studies

6 Food enzymes

SIBEL ROLLER and PETER W. GOODENOUGH

6.1 Introduction and historical perspective

Although more than 2300 enzymes have been catalogued by the International Union of Biochemistry and Molecular Biology (Webb, 1992) and several thousand more have been described in the scientific literature, only about 20 enzyme types are sold in large enough quantities for industrial-scale processing. Of these, only about 8 are used in quantities of more than 10s of kilograms. Nevertheless, the use of industrial enzymes as food processing aids has grown into a multi-million dollar industry over the last 30 years, as shown in Table 6.1.

The food industry is the second largest user of enzymes (the detergent industry is first). The US market alone for food processing enzymes was estimated at $168.8 million in 1996 and is expected to grow to around $186 and $214 million by the years 2000 and 2006, respectively (Wrotnowski, 1997). In 1995, the value of the total worldwide enzyme market (including food and detergent applications) was estimated at $1 billion and was projected to grow to $1.7 billion by the year 2005 (Godfrey and West, 1996).

The reduced cost and increased availability of industrial enzymes has been made possible by tremendous progress made since the 1950s in the development of the submerged fermentation technique. Although some animal and plant enzymes such as rennet and papain are still in use, the large majority of bulk enzymes are today manufactured using fungal or bacterial fermentation. In the case of one enzyme, glucose isomerase, advances in the technology of immobilization made in the early 1970s have also been important in enabling the development of a commercially viable industrial process for the production of high-fructose corn syrup (HFCS) from starch; this technique allowed for much greater utilization of the relatively expensive enzyme by fixation of the protein onto a column through which a continuous stream of the liquefied starch slurry could be passed. In the late 1980s, another new and very important development took place: the first food enzymes derived from genetically modified organisms (GMOs) entered the markets.

In this chapter, we examine the development of genetic approaches to the manufacture of enzymes as food processing aids. The theme is developed initially by a description of how the introduction of several innovative

Table 6.1 Common food enzymes, their applications and worldwide sales

Enzyme group	Common name	Substrates	Principal applications	World sales in 1990[1] (millions of US $)
Proteases	chymosin (animal and microbial)	milk protein	milk coagulation in cheese manufacture	75
	papain	vegetable proteins, meat	protein hydrolyzates, yeast extracts, meat tenderization	8
	trypsin	vegetable proteins, meat	protein hydrolyzates	8
Carbohydrases	amyloglucosidase	starch	sweetening syrups	75
	α-amylase	starch	sweetening syrups, maltodextrins	50
	invertase	sucrose	soft-centred chocolates	8
	pectinase	pectin	fruit juices	7
Isomerases	glucose isomerase	glucose	high fructose corn syrup (HFCS)	40
Others	β-glucanase	glucans	wine	20
	cellulase	cellulose	fruit juices	
	dextranase	dextran	sugar beet processing	
	glucose oxidase	glucose	treatment of egg whites	
	hemicellulase	hemicellulose	fruit juices, bread	
	lactase	lactose	hydrolysis of lactose	
	lipase	plant lipids	interesterification	
	pullulanase	starch	HFCS	

[1] Sales figures from: Olsen, H-S. (1995) Enzymes in Food Processing, in *Biotechnology*, Vol. 9, *Enzymes, Biomass, Food and Feed*. G. Reed and T. W. Nagodawithana (eds). VCH Weinheim. pp. 663–736; reproduced with permission.

recombinant sources of chymosin has solved a long-standing supply problem in the dairy industry. This is followed by a brief overview of the use of GMOs as novel sources of other food enzymes now and in the future. Finally, the chapter concludes with a description of protein engineering and how knowledge of the detailed, three-dimensional structure can drive the modification of proteins so that improved enzymes can be developed for use in industry.

6.2 Chymosin

6.2.1 The need for a milk-clotting enzyme

Chymosin is the most important enzyme in the dairy industry and is used to clot milk during the manufacture of cheese. Traditionally, chymosin was obtained from rennet (also containing pepsin) extracted from the fourth stomach of young calves. However, the continuing decline in the slaughter of young calves and the increasing rate of cheese production worldwide (estimated at over 14 million tonnes per year and requiring over 50 000 kg of chymosin), has led to an increasing gap between supply and demand for high quality rennet (Roller *et al.*, 1994).

Several alternative sources of milk-clotting enzymes exist but none of these has been as widely applicable as calf rennet. Only six rennet substitutes have been almost acceptable for one or more cheese varieties. These are the pepsins from cow, pig and chicken and the fungal acid proteases from *Mucor miehei*, *Mucor pusillus* and *Endothia parasitica*. Chicken pepsin has been a poor substitute for chymosin and is only used in Israel and the former Czechoslovakia. Bovine pepsin is probably the best of the substitutes. Although the initial specificity of bovine pepsin is similar to that of calf chymosin, there is a greater amount of non-specific proteolytic activity which weakens the milk protein network and leads to decreased yields of curds. Microbial proteases, available since the 1970s, have been used successfully to prepare some young cheeses, but again, their non-specific action renders them less suitable for the production of matured cheese due to the formation of undesirable flavours and odours.

6.2.2 *Milk clotting*

Milk is a complex biological fluid which contains a large quantity of nutritionally important proteins but also water, fat, lactose, citric acid and inorganic components including calcium phosphate. Most of the proteins in milk are caseins which have four different sequences and are known as α-s1, α-s2, β and κ-caseins. These molecules make up about 80% of the total milk protein – in all about 27 g/l. Casein molecules have a less well-defined tertiary structure than the globular whey proteins lactalbumin and lactoglobulin, also present in milk (Kumosinski and Farrell, 1994). The α-casein and β-casein associate into compact micelles with a well-defined structure with κ-casein on the micellar surface. Micelles are highly hydrated, roughly spherical with a diameter of 100 nm, contain about five thousand casein monomers and scatter light giving the traditional 'milky appearance'.

Chymosin is the major aspartic acid protease used to cleave the κ-casein after acidification of the milk by lactic acid bacteria starter cultures. Cleavage of the κ-casein protein occurs between the amino acids phenylalanine (number 105 in the protein chain) and methionine (number 106). The remaining casein chains aggregate and, instead of forming micelles, precipitate to form a solid protein mass. Clotting takes about 20 min at pH 6.5 and 30°C, after which the curd mass is cut, drained and pressed to remove the whey proteins and water. The curds continue to 'mature' with time and during the maturation phase some protease activity continues.

6.2.3 *The structure and synthesis of animal chymosin*

Aspartic acid proteases are a large family of enzymes with very important roles in many processes. Chymosin is also known as rennin and is the main enzymatic component of calf rennet. At an early stage of synthesis, the enzyme is formed as pre-prochymosin. As it is secreted into the stomach, the pre-region is removed to produce prochymosin which comprises 365 amino acids. Prochymosin is catalytically inactive and is autocatalytically cleaved to one of two active species. At pH 4.2, 42 N-terminal residues are removed to produce active chymosin of MW 35 600. At pH 2.0, only 27 amino acids are removed producing active pseudo chymosin. Pseudo chymosin has a MW of 37400 and is stable below pH 3.0 or above pH 6.0 but is further converted to chymosin by incubation at pH 4.5. Although the activity of calf chymosin is relatively low against proteins at the near-neutral pH used in cheese-making, the specific activity against caseins is just sufficient for the process to provide the right balance of taste and texture in cheese.

Chymosin from calf stomach has a kidney shaped structure with approximate overall dimensions of 40 Å x 50 Å x 65 Å (Figure 6.1). There are two domains in the enzyme: domain one runs from residues 1–175 and domain two comprises residues 176–323. Secondary structure consists primarily of parallel and antiparallel β strands in each domain. The domains are separated by a long cleft that contains the amino acid residues involved in catalysis – these are aspartic acids 34 and 216. Such a cleft is a feature often found in the active sites of enzymes that cleave polymeric chains (Schecter and Berger, 1967). In rennin, the active site cleft is unusually long and recognizes the sequence of amino acids histidine 98-lysine 111 in K-casein. A flap covers the two domains and this loop of polypeptide is made up of residues 77–86. Chymosin has some unique features in this loop area that seem to be important to its mechanism of action. In particular, tyrosine 77 of the flap region does not have a hydrogen bond to nearby tryptophan 42 as would be expected but protrudes from the pocket to interact hydrophobically with phenylalanine 119 and leucine 32.

6.2.4 *The genetic modification of microorganisms for chymosin production*

In the early 1980s, recombinant DNA techniques were developed which made it possible to transfer DNA from one species (e.g. an animal) to another (e.g. a microorganism). This revolutionary new technology and the growing market need for a standardized chymosin of high quality with a constant and guaranteed supply and stable price combined to lead to the eventual solution of the chymosin supply problem in the dairy industry. The

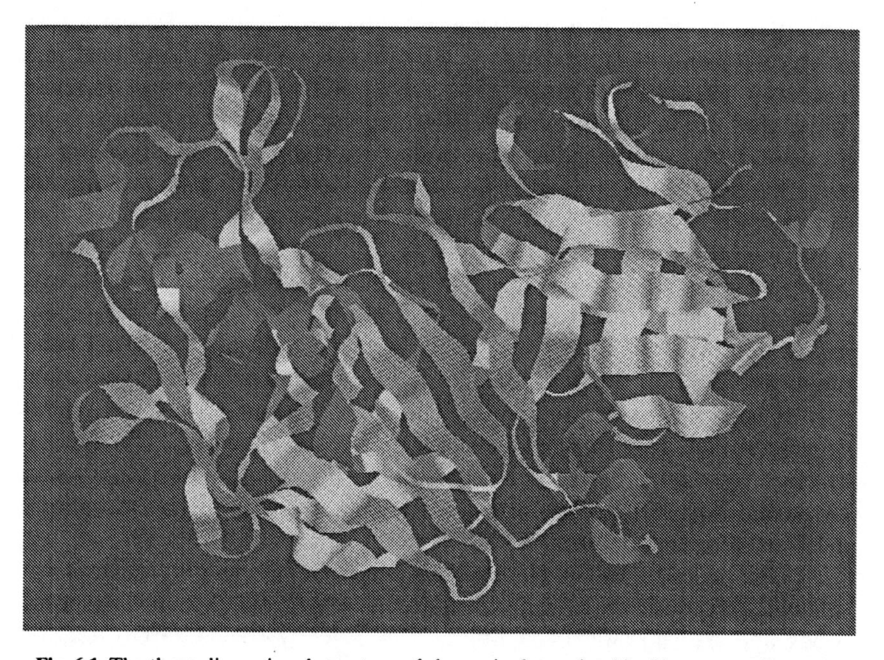

Fig. 6.1 The three-dimensional structure of chymosin determined by X-ray crystallography.

chymosin story has been a prime example of how genetic modification can be used to bring benefits to the food processing industry and ultimately to the consumer. The DNA of calf chymosin has been successfully cloned (copied) into yeast (*Kluyveromyces lactis*), bacteria (*Escherichia coli*) and moulds (*Aspergillus niger* var. *awamori*) to give at least three commercial products which are now used widely in cheese manufacture.

Several commercial organizations, including Pfizer, Gist-brocades and Genencor International, undertook the development of chymosin from genetically modified microorganisms. At Pfizer's, scientists chose to work with *E. coli*, as the genetics and biochemistry of this organism were well-understood and several genetic constructs were already in existence. In early work, Emtage and colleagues (1983) reconstructed the chymosin gene *in vitro* from a series of enzyme fragments and copied it into *E. coli* using a plasmid. Insertion was immediately downstream of a strong *E. coli* trp promoter and a functional ribosome binding site. This meant that the recombinant chymosin had a methionine residue (not the normal N-terminal amino acid) as the first amino acid. Consequently, further processing steps were necessary to produce authentic chymosin (Emtage *et al.*, 1983; Kawaguchi *et al.*, 1987; McCaman *et al.*, 1985; Marston *et al.*, 1984).

One of the potential disadvantages of recombinant expression of a protein is that the material may be produced in a form that is not normal (native). Many eukaryotic proteins have difficulty in folding into the correct tertiary form when they are expressed by prokaryotes. In particular, eukaryotic proteins expressed in *E. coli* are often actively partitioned into a separate part of the organism. The partitioned proteins are not correctly folded and the insoluble mass of heterologous protein accumulates in an inclusion body. Although inclusion bodies are considered an advantage by some workers as they provide a concentrated amount of pure protein that can be easily separated by centrifugation from the other bacterial cell contents (Marston *et al.*, 1984), this advantage is outweighed by the problem of obtaining correctly-folded active enzyme following extraction from the cell.

Another drawback of using *E. coli* to produce chymosin is that this organism is associated with poor hygiene in the food industry and is not commonly accepted for the production of food ingredients due to the potential risk of bacterial toxin production. Therefore, several food-grade microorganisms including *Bacillus licheniformis*, *Lactococcus* spp. and *Saccharomyces cerevisiae* have been assessed for production of recombinant chymosin. However, problems with secretion of the enzyme into the culture medium were encountered.

At Gist-brocades in The Netherlands, the yeast *Kluyveromyces lactis* was selected for genetic modification as it occurred naturally in many fermented foods and had already been affirmed as safe by the US FDA for the production of the enzyme lactase for food processing (FDA, 1984). In addition, a plasmid carrying the prochymosin DNA had become available for transfer into *K. lactis* (Maat *et al.*, 1983). The cloning (copying) of the entire calf genomic DNA (containing numerous sequences of 'junk DNA' not needed for chymosin synthesis) was avoided by isolating and purifying chymosin messenger RNA (mRNA) using the enzyme reverse transcriptase (Harris *et al.*, 1982; Maat *et al.*, 1983). This technique was an important technical breakthrough at the time because it allowed genetic modifications to be made without full knowledge of the amino acid sequence of the target protein. More recently, alternative advanced techniques such as the polymerase chain reaction (PCR) have been developed (see Chapter 1); however, the efficient use of PCR relies on full knowledge of the amino acid sequence of the target protein.

The expression of genes located on plasmids can be very unstable. Consequently, it is desirable to integrate new genes into the chromosome of the host. Such a strategy was adopted in the genetic modification of *K. lactis*. A lactase promoter (start signal) was added in front of the chymosin sequence and a lactase terminator (stop signal) was attached at the end. The chymosin gene was inserted into the host yeast following treatment with lithium salts and transformants were selected by detection of resistance to kanamycin. The amount of prochymosin produced by the transformed cells was quantified using a milk-clotting assay (van den Berg *et al.*, 1990; Sambrook *et al.*, 1989). Up to 95% of the prochymosin produced by the transformed yeasts was found extracellularly in the culture medium. Integration of the new gene into the chromosome was successfully achieved and remained unchanged for at least 100 generations (van den Berg *et al.*, 1990).

In addition to *E. coli* and *K. lactis* developed at Pfizer and Gist-brocades, respectively, the third major player in recombinant chymosin has been Genencor International with *Aspergillus niger* var. *awamori* as the host organism. All three enzymes are produced commercially under the trade names Chy-Max® (Pfizer), Maxiren® (Gist-brocades) and Chymogen® (Genencor). In 1996, the entire Chy-Max® product line including patents and associated intellectual property rights was acquired from Pfizer by Christian Hansen's Inc, a Danish manufacturer of starter cultures, enzymes and food ingredients (Anon., 1996).

6.2.5 The industrial production of recombinant chymosins

All three recombinant chymosins currently available on the market are produced by standard large-scale fermentation methods. The chymosin from the genetically modified *E. coli* is extracted using a series of steps which include: separation of the cell mass from the fermentation broth by centrifugation, washing of the cell pellet and suspension in a small volume of buffer, disruption of the cells, separation of the inclusion bodies containing the enzyme from the rest of the cell debris by centrifugation, washing of the centrifuged inclusion bodies with acid to destroy any remaining whole cells, solubilization of the inclusion bodies by alkaline urea, and purification with anion exchange chromatography (Flamm, 1991).

In the case of prochymosin production by the genetically modified *K. lactis*, fermentation is stopped by the addition of benzoate to kill the yeast cells. In addition, sulphuric acid is added to reduce the pH to 2 which has the dual function of inactivating proteolytic enzymes other than chymosin and inducing the conversion of prochymosin to chymosin, as is the case in the calf stomach. The fermentation broth is then filtered to remove yeast cells and concentrated using ultrafiltration.

6.2.6 Technological performance of recombinant chymosins

Numerous studies have demonstrated that cheese produced using recombinant chymosin was essentially indistinguishable from traditional cheese (Barbano and Rasmussen, 1991; van den Berg, 1991; Bines *et al.*, 1989; Broome and Hickey, 1990; Green *et al.*, 1985; Hicks *et al.*, 1988; Roller *et al.*, 1994). For example, the recombinant enzyme from *E. coli* has been compared with conventional chymosin in cheeses prepared from 90-litre batches of milk (Green *et al.*, 1985). After pressing, cheeses were vacuum-packed and ripened at 12–13°C. Samples were tested for gross composition, including moisture, starter counts, concentration of caseins and nitrogen content in both curds and cheeses. Firmness and elasticity were also measured. Triplicate analysis showed that equal amounts of the conventional and recombinant rennets had the same activity in milk coagulation. There was no measurable difference between the degradation of the different subtypes of caseins in cheeses made with recombinant or standard rennet. Cheese made with recombinant rennet tended to be marginally firmer than the controls but this was not highly significant. No differences in long-term rates of proteolysis were detected (Green *et al.*, 1985).

Large-scale, independent trials of recombinant chymosin from *K. lactis* have been carried out in several varieties of cheese including Appenzell, Camembert, Cheddar, Edam, Emmental, Gouda, Hispanico, Italico, Kashar, Manchego, St Paulin, Tilsit and White Pickled in several countries (Spain, Germany, France, USA, Turkey, Italy, The Netherlands, Austria, Finland, Ireland, Switzerland and the UK). In these trials, milk coagulation, curd firmness, ageing profiles, flavour development and organoleptic qualities have been reported to be identical for both the recombinant and animal sources of chymosin (Roller *et al.*, 1994).

6.2.7 Health, safety and environmental issues

Purified chymosin from genetically modified *K. lactis* has been shown to be structurally identical to the animal-derived enzyme by several methods. The molecular size and weight, determined by high performance size exclusion chromatography and polyacrylamide gel electrophoresis, have been found to be the same. The native and recombinant chymosins reacted in the same way when tested using monoclonal antibodies and Western immunoblotting (Sambrook *et al.*, 1989). Indeed, X-ray crystallography has now shown that every atom in the recombinant enzyme is in a similar position to that in the native enzyme. Furthermore, the activity of native and recombinant chymosins has also been demonstrated to be essentially the same in terms of temperature, pH and calcium sensitivity. Very similar work has been carried out to demonstrate that the other two recombinant chymosins from bacterial and fungal sources were also substantially equivalent to the animal-derived enzyme (Emtage *et al.*, 1983; Gustchina *et al.*, 1996; Kawaguchi *et al.*, 1987; McCaman *et al.*, 1985; Marston *et al.*, 1984).

The recombinant chymosin preparations are substantially purer than traditional rennet. Whereas the material extracted from calf stomachs is usually brown and contains about 2% chymosin, the recombinant preparations are clear and may contain as much as 80% protein (Flamm, 1991). It has been calculated that the use of recombinant chymosin in food processing could result in an average daily intake of less than 0.5 mg per person or 10 μg per kg body weight (Flamm, 1991).

All of the compositional studies described above have indicated that the chymosins from GMOs would be safe for use in human foods. Nevertheless, since these were the first products of recombinant DNA technology destined for human food, additional precautionary measures were taken to ensure their safety. For example, toxicological trials carried out in the USA have shown that single doses as high as 5 g of the *K. lactis* chymosin/kg of animal body weight did not produce any signs of oral toxicity in rats.

No evidence of any mutagenic tendencies was detected in three different *in vitro* mutagenicity tests nor in animals fed large amounts of recombinant chymosin. In a short-term toxicity study, rats fed on cheese prepared with recombinant chymosin from *K. lactis* showed no evidence of systemic toxicity. No abnormalities were found in a 91-day subchronic toxicity study on rats fed with the recombinant chymosin alone as well as with cheese prepared using the recombinant source of the enzyme. Notably, allergenicity tests showed that recombinant chymosin from *K. lactis* was less of a sensitizer of guinea pig skin than the traditional animal source of chymosin (Roller *et al.*, 1994). Presumably, this was because the recombinant source was purer than the animal material.

Similarly, in the case of the chymosin from recombinant *E. coli*, steps have been taken and studies carried out in an attempt to eliminate as many safety concerns as possible. Since the production strain carried an ampicillin resistance marker, Pfizer's added an acidification step to the extraction procedure to ensure that no functional copies of the gene were present in the final product. Furthermore, it has been demonstrated that there was no detectable transformable DNA in the Pfizer enzyme preparation and no DNA fragments larger than 200 bases detectable by radiolabelled hybridization after gel electrophoresis. By comparison, the coding sequence for the antibiotic resistance gene carried by the producer strain contained 858 bases (Sutcliffe, 1978). Previous published studies had shown that *E. coli* K12 was not a pathogenic organism and did not colonize the human gut even if consumed at a level of 10^{10} viable organisms per ingestion. Furthermore, it was argued that non-pathogenic strains of *E. coli* are part of the normal gastrointestinal flora of man where they are found at a level of 10^6–10^8 organisms per g (Curtiss, 1978; Gorbach, 1978; Smith, 1978). Finally, in their petition to the FDA, Pfizer's provided results of an *in vitro* assay which showed that their strain produced no shiga-like toxin, and two *in vivo* studies (a 5-day study in dogs and a one-month study in rats) showing no adverse effects on feeding the recombinant chymosin at a range of doses (Flamm, 1991; Newland and Neill, 1988). Therefore, although *E. coli* was not a common food-grade organism, it has been concluded that the organism was safe for producing chymosin (FDA, 1990).

Monitoring of environmental contamination around a French production site producing chymosin from recombinant *K. lactis* has been reported. Although some live material (700 cfu (colony forming units) per cubic metre of air) was vented into the atmosphere, there was no evidence of survival and growth of the yeast in soil surrounding the site. This observation was not unexpected as most industrial strains used in commercial fermentations are not well-suited for survival in the environment. Water contamination with live recombinant yeast was thought to be obviated by routine treatment of the fermentation broth with sodium benzoate and sulphuric acid to inactivate the cells (Roller *et al.*, 1994).

6.2.8 *Legislative and labelling position*

All three recombinant chymosins described in previous sections are now permitted for use in foods in the USA and in many European countries. The FDA received the first petition requesting GRAS (generally recognized as safe) affirmation for a recombinant chymosin in 1987 from Pfizer and approved it in 1990 (FDA, 1990; Flamm, 1991). The Gist-brocades product has also been given GRAS status by the FDA in 1992, as well as receiving clearance by the Scientific Committee for Foods of the EC (Praaning-van Dalen, 1992). The Genencor product was approved by the FDA in 1993 (Anon., 1993).

The use of enzymes as processing aids in foods does not require labelling. However, some retailers have taken the inititative of explaining the consumer benefits of using such enzymes on selected food labels. For example, the Co-operative Retail Society in the UK is selling vegetarian Cheddar cheese made with chymosin from GMOs with the label: 'Produced using gene technology and so free from animal rennet'.

6.2.9 *Consumer acceptance and market penetration*

The marketing of recombinant chymosins was first targeted at cheese manufacturers who constituted the first line of 'consumers' for these new products. The very clear benefits of consistency and constancy of supply have led to the widespread adoption of the recombinant chymosins as substitutes for animal-derived rennet. In 1992, the recombinant chymosins were estimated to command approximately 25% and up to 70% of the cheese-making market in European countries (where permitted) and in the USA, respectively (Guinee and Wilkinson, 1992). In the first three years since receiving regulatory approval, the Pfizer product Chy-max® had reportedly been used to make more than 15 billion pounds of cheese in 17 countries (Scher, 1993a,b).

Since the use of enzymes in food processing does not require labelling, it is doubtful whether many consumers are aware that chymosin or any other enzymes, whether native or from GMOs, are being used in the production of cheeses or any other foods. Nevertheless, some retailers have voluntarily labelled foods prepared with the aid of recombinant enzymes, as noted in section 6.2.8.

6.3 Other food enzymes from GMOs

It has been estimated by some industrial observers that 50% of all industrial enzymes (including food and detergent enzymes) are already being manufactured using GMOs (Dalboge, 1995). For example, most lipases used in many cold-wash detergents are produced by GMOs, and phytase, a plant-derived enzyme that has been engineered into *Aspergillus niger* is reportedly used in 50% of all pig feed in The Netherlands (Hodgson, 1994). In the food industry, however, uptake may be somewhat slower, probably because of regulatory hurdles (section 6.3.2). Some examples of food enzymes produced commercially (or very soon to be produced) with the aid of GMOs are shown in Table 6.2.

The development of enzyme production from GMOs has been driven by three major forces: the need by the enzyme supplier to improve production efficiency (i.e. enzyme yields from fermentation) and to maintain profit margins in the face of declining prices; the need to improve enzyme performance in existing applications; and the demand for enzymes suitable for novel applications at an economic price.

6.3.1 Recombinant enzymes for high fructose corn syrup production

The starch conversion industry is the second largest enzyme-consuming industry after the detergent industry. By far the biggest product is high fructose corn syrup (HFCS), produced principally using alpha-amylase, amyloglucosidase and glucose isomerase. Two grades of HFCS containing 42 and 55% fructose account for over a third of the caloric sweetener market in the USA. Annually, more than 8 million tons of HFCS are produced (Olsen, 1995).

Although the HFCS process is now well-established, processors and enzyme manufacturers are constantly seeking new ways of increasing efficiency by reducing processing time and simplifying processing procedures. Until recently, the processes of liquefaction, saccharification and isomerization had to be carried out sequentially using alpha-amylase (from *Bacillus licheniformis*), amyloglucosidase (from *Aspergillus awamori*) and glucose isomerase (from *Actioplanes moissouriensis*) at temperature/pH regimes of 105°C/pH 6, 60°C/pH 4.5 and 60°C/pH 7.5, respectively. These conditions were a compromise between the physical properties of the starch substrate and the temperature, pH and cation requirements of the different enzymes. Clearly, it would be more cost effective and less labour-intensive to carry out the entire process in a single vessel under one temperature and pH regime.

In the 1980s, the first petitions to the FDA for GRAS affirmation of amylase enzymes from recombinant strains of *Bacillus subtilis* were submitted by CPC International (a major producer of HFCS) and its enzyme

Table 6.2 Examples of commercial or near-market food enzymes from GMOs

Enzyme	Application	Source organism	Status
Acetolactate decarboxylase	Accelerating beer maturation by reducing diacetyl levels	GM *Bacillus*	Commercially available in some countries; undergoing approval process in UK
Maltogenic alpha-amylase	Anti-staling in bread	Gene from *B. stearothermophilus* cloned into *B. subtilis*	Commercially available from Novo Nordisk
Xylanase	Improving dough processability, crumb structure and loaf volume in bread	GM *A. oryzae*	Undergoing approval process in UK
Hemicellulases	Improving dough processability, crumb structure and loaf volume in bread	GM *B. subtilis* (Rohm GmBh) GM *A. niger* var. *awamori* (Quest International) – both self-cloned to improve yields	Approved by UK ACNFP in 1996
Lipase	Interesterification of palm oil to produce cocoa butter substitutes	GM *A. oryzae*	Developed by Unilever (Loders Croklaan)
Cyclomaltodextrin glycosyltransferase (CTG-ase)	Production of ring-like cyclodextrins for flavour and aroma binding	Gene cloned from *Thermoanaerobacter* spp. to *Bacillus* to make high temperature, low pH industrial processing possible	Under development by Novo Nordisk

Compiled from: ACNFP (1995, 1994); CCFRA (1997); Dornenburg and Lang-Hinrichs, 1994; Pedersen *et al.* (1995); MAFF COT (1994).

subsidiary Enzyme Biosystems Ltd. (CPC, 1984; EBS, 1986). The gene coding for an alpha-amylase from *Bacillus stearothermophilus* had been transferred into *B. subtilis* to give an enzyme with better thermal and pH stability, as well as better stability in the presence of very little or no calcium, than the conventional enzyme (Tamuri *et al.*, 1981). The latter quality was particularly attractive because calcium adversely affected the other two enzymes used in HFCS production, amyloglucosidase and glucose isomerase. Another gene, coding for an amylase from *Bacillus megaterium*, was also cloned (copied) into *B. subtilis* (Marie-Henrietta *et al.*, 1985). The new enzyme was found to offer better economies when used with amyloglucosidase during saccharification because it could be used at higher solids levels, and it resulted in increased glucose yields, reduced reaction times, and reduced concentrations of oligomers in the product (Sharma, 1989).

Although it was reported in 1989 that these two recombinant amylases were being used in commerce (Arbige and Pticher, 1989), quantitative information on the extent of use (e.g. by weight or by value) is not readily available.

6.3.2 *Safety, legislation and labelling*

The majority of food enzymes are used as processing aids and have no function in the final food product. Consequently, enzymes are not true additives and do not need to be declared on the label in many countries. A common exception to this has been the use of lysozyme as a cheese preservative; in Europe, lysozyme must be declared on the label using the appropriate class name, E-number and generic name.

In the absence of a positive list of enzymes with defined limits on the amounts which may be used in specific foods or of a negative list prohibiting specific enzymes, the use of enzymes in the UK is subject to the control of the general provisions of the Food Safety Act 1990. The Food Safety Act stipulates the need for food, and substances added to food, to be safe and to be composed and presented in such a way that purchasers are not misled as to its nature, substance or quality. It would be an offence to use an enzyme if this made the food harmful to the consumer.

In the European Union, common regulations on food enzymes have not yet been developed. The Association of Microbial Food Enzyme Producers (AMFEP) has played an important role in promoting harmonization and in 1996 has proposed to the European Commission that a single set of regulations for food enzymes, whether from wild-type or genetically modified organisms, should be developed. Until such time as harmonization of regulations across Europe and worldwide is achieved, national regulations still need to be consulted and these may vary widely from country to country. For example, in The Netherlands, approval is required for enzymes from GMOs only while in France and Spain, enzymes are considered as additives and authorization is required for all such preparations, whether derived from GMOs or not (AMFEP, 1996; Hamer, 1995).

In the USA, the FDA regulates food enzymes under the Food, Drug and Cosmetic Act (FDCA). This states, in brief, that anything that is added to a food or affects the characteristics of a food is a food additive unless it is GRAS or approved for use prior to 1958 (Fordham and Block, 1987). The GRAS petition accepted by the FDA from Pfizer for chymosin from *E. coli* suggests that enzymes produced by organisms which would not normally be considered 'food-grade' may be affirmed as GRAS, provided sufficient information is available to assess their safety, as discussed in more detail below (FDA, 1990).

The worldwide regulatory picture for food enzymes is further complicated by compositional regulations in different countries. For example, in the UK, there are regulations which set out the ingredients which may be used to manufacture foods such as cheese, bread and chocolate. Until recently, these regulations meant that the use of enzymes in these foods was rather limited. For example, the Bread and Flour Regulations 1995 stated that only alpha-amylases, proteases and hemicellulases may be added to flour during breadmaking. However, with effect from 1 July 1996, all restrictions on the use of enzyme preparations in bread, flour and cheese were removed as part of a continuing deregulation programme by the UK Government. The relaxation of compositional regulations in the UK will allow for increased usage of enzymes in these foods in the late 1990s and beyond. However, the use of enzymes in baking is still strictly regulated in many other countries. For example, France and The Netherlands have a 'positive list' of enzymes which may be used in baking (Hamer, 1995).

Microbial sources of food enzymes must be non-pathogenic and non-toxigenic. Over the years, several species of fungi (e.g. *Aspergillus niger* and *A. oryzae*) and bacteria (e.g. *Bacillus subtilis* and *B. licheniformis*) have acquired a record of safe use as sources of a wide range of food enzymes. For enzymes from recombinant organisms, the primary regulatory status is determined by the host organism, the donor organism, and the vectors used to achieve genetic transfer (AMFEP, 1996).

Specifications and guidelines for the safety assessment of microbial enzyme preparations used in food have been developed by several bodies at the national (Committee on Toxicity, UK, 1992) and international level (AMFEP, 1996; Scientific Committee for Food (Europe), 1991; FDA's 'Red Book', 1982; Joint FAO/WHO Expert Committee on Food Additives (JECFA), 1984 and 1987; Food Chemicals Codex, 1981; Fordham and Block, 1987; Pariza and Foster, 1983). These guidelines set out criteria which must be met by all food grade enzymes and include evaluation of the raw materials used in the microbial fermentation, determination of chemical contaminants, microbial contamination, mycotoxins in fungal enzymes and antibiotic activity in the final products (Pedersen *et al.*, 1995). Whenever necessary, the safety of every enzyme preparation, whether from a GMO or not, is further demonstrated on a case-by-case basis by appropriate feeding studies. In the UK, acceptance of a novel food enzyme by the independent Food Advisory Committee (FAC) appointed by the Government is dependent not only on safety data but also on a proven case of need for the new substance.

6.4 Protein engineering of enzymes

Some of the disadvantages of enzymes, such as relatively poor stability at high processing temperatures, are being addressed by the latest developments in protein engineering. The 'engineering' of a protein essentially involves changing one or more amino acids in the primary sequence in order to change the folding of the three-dimensional (tertiary) structure of the protein. Subtle changes in tertiary structure (e.g. changes in hydrophobic interactions, hydrogen bonding, electrostatic interactions and disulphide bridging) can have profound effects on the activity and stability of enzymes.

There are two main prerequisites for successful protein engineering. Firstly, the three-dimensional structure of the native protein must be known so that its properties can be related to structure. Secondly, the gene coding for the protein must be known so that it can be cloned (copied) from the original wild type genome into a vector, and used to make very specific site-directed mutations. Although many gene sequences are now available, the tertiary structures of many proteins, especially industrial food enzymes, are still not known. In spite of the availability of sophisticated computer modelling packages, it is still not possible to accurately predict the tertiary structure of proteins from the primary amino acid sequence alone; consequently, expensive and lengthy experimental techniques such as X-ray crystallography on protein crystals and nuclear magnetic resonance on proteins in solution still need to be undertaken before a protein can be engineered. However, it has been estimated that by the year 2000, tertiary structures of most proteins in microorganisms will be known allowing for a much faster rate of progress in the engineering of industrial enzymes (Goodenough, 1995). The first example of successful protein engineering of an industrial food enzyme has been glucose isomerase, described in more detail in section 6.4.3.3.

6.4.1 The structure and function of proteins

All enzymes are proteins or polypeptides with four levels of structure. The primary structure is the amino acid sequence. This sequence predisposes the molecule to form the secondary structure which occurs in the shape of random coils, helices or β-pleated sheets. Interactions between amino acids (e.g. cysteine) by disulphide bridges, electrostatic interactions (e.g. opposite charges attract) and hydrophobic interactions wrap the secondary structure of an enzyme into the complex tertiary or three-dimensional structure. When the protein interacts with other proteins, a quaternary structure is formed, usually by hydrophobic or electrostatic interactions.

In addition to the protein moiety, some enzymes have non-protein components. For example, some industrial food enzymes require metal ions as cofactors for catalysis. These may be loosely or tightly bound to the protein by covalent or non-covalent bonds. Cofactors are often very important components of enzymes without which activity and/or stability can be reduced or completely eliminated.

Proteins are almost unique amongst molecules in the diversity of their three-dimensional forms (Wulthrich and Hendrickson, 1996). The first three-dimensional structure of an enzyme (lysozyme) was solved in 1966 (Phillips, 1966) but it was not until the 1980s that a large number of additional protein structures became available. Structurally, proteins could be divided into three groups: those which have all of their tertiary structures built from β-sheets, those with tertiary structures made entirely of α-helices, and those with a mixture of the two motifs.

While the overall structure of a protein may be unique for a particular function, certain basic tertiary folds can be found in many proteins with widely differing functions. In the case where there is a common tertiary fold, the unique part of any structure is usually in the loops of protein which connect the conserved secondary structural elements. It has been suggested that there are one to two thousand basic folds conserved in nature (Chothia, 1988, 1992) but recent evidence from the first complete genome sequence of yeast suggests that the number may be much higher (Goffeau *et al.*, 1996).

One of the first basic motifs to be discovered was the eight-fold α/β barrel of triosephosphate isomerase (commonly referred to as the TIM barrel). In this structure, the protein chain can be traced from sheet to helix eight times and the molecule forms a rounded structure, hence the descriptor 'barrel'. The loops between the secondary structure are the sites of binding of many different substrates except for narbonin. Narbonin does not bind any other molecules and its function is as a plant storage protein. In the 30 other examples of this motif so far described, the function of the molecule is diverse. For example, many TIM barrels bind and catalyse the hydrolysis of a range of carbohydrates (Jenkins *et al.*, 1995) (Figure 6.2).

In contrast to the common barrel motif, there are rare protein folds which have only been found a small number of times. An example of this has been a protein formed of a parallel β sheet structure. Previous hypotheses had predicted that a parallel β sheet structure would not be stable when it was unexpectedly discovered in certain pectic enzymes (Figure 6.3). Although it is now known how the internal conservation of residues stabilizes the structure by internal hydrogen bonding, it was not possible to predict such a structure prior to obtaining experimental data. Initially restricted to those enzymes which bind and hydrolyse pectate or pectin (polygalacturonate or the methylester of polygalacturonate) this motif has now been found in a protease and a protein from viruses (Pickersgill *et al.*, 1994; Yoder *et al.*, 1993).

By combining all available data on structures with advanced computer manipulations, it is possible to predict the effect of single amino acid alterations on the tertiary structure of proteins. The strengths of electrostatic interactions between negatively charged amino acid residues (glutamate and aspartate) and positively charged residues (arginine, lysine and histidine) can be quantified (Warwicker and Watson, 1982; Gilson and Honig,

Fig. 6.2 A xylanase with a classical eight-fold β/α structure found over 30 times in natural protein folds. The strands of protein progress from helix to sheet eight times forming a barrel structure. (Determined by Harris *et al.*, 1995, reproduced with permission.)

1987). Lengths of hydrogen bonds can be calculated and positional problems, such as hydrophobic residues that are partially solvent exposed, can be predicted by inspecting the tertiary structure on a molecular graphics screen.

6.4.2 Site-directed mutagenesis

The technique of site-directed mutagenesis was pioneered in the early 1980s by Smith who received a Nobel prize for this work in 1993 (Smith *et al.*, 1985; Zoller and Smith, 1982). Before site-directed mutagenesis can be undertaken, the gene coding for the target protein must be known. Particular oligonucleotides are then chemically synthesized so that they hybridize with a specific complementary base sequence of a section of native DNA *in vitro*. The synthetic oligonucleotide usually differs from the natural DNA in one, two or three contiguous nucleotides in such a way that an individual amino acid in the target protein is changed to any one of the other naturally occurring amino acids. Once a single amino acid has been changed, the mutant DNA is amplified and selected. This can be done in a number of ways but some genetic tricks can be used to 'identify' the mutant DNA and destroy the recombinant parental DNA *in vivo*. This includes the use of bacteria as

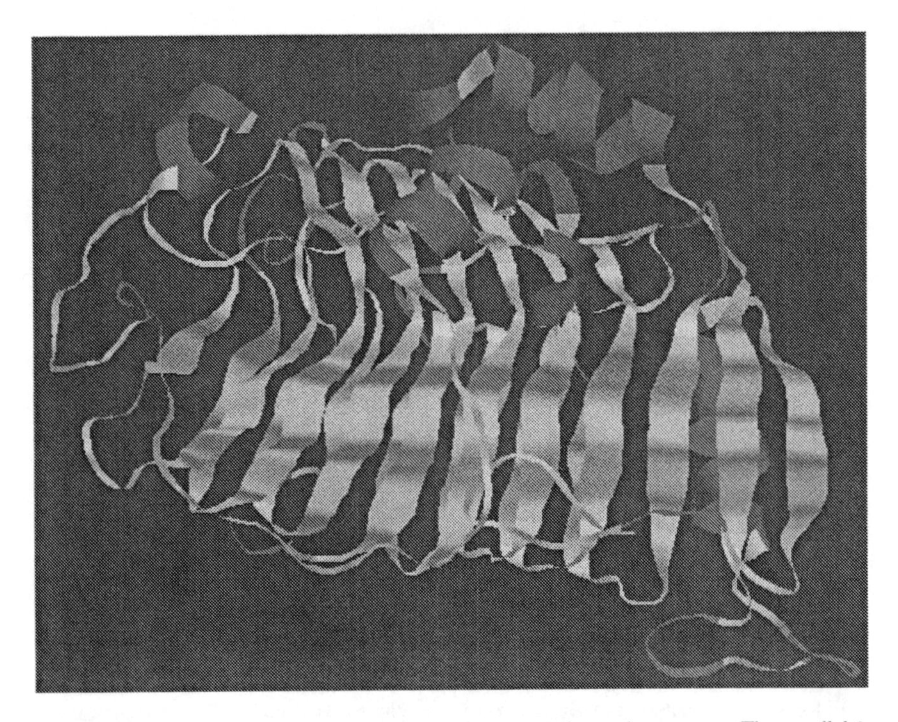

Fig. 6.3 The three-dimensional representation of an unusual protein structure. The parallel β sheet motif is shown here in pectate lyase (Pickersgill *et al.*, 1994, reproduced with permission).

hosts which are deficient in some enzymes which in turn allows deoxyuridine to be incorporated into DNA in place of thymidine. If the mutant DNA strand has thymidine in positions where the parental DNA strand has uridine, and the DNA is transferred to bacteria with a normal complement of enzymes, the parental DNA is degraded and colonies of bacteria with only mutant DNA can be selected.

6.4.3 Improving the properties of enzymes by protein engineering

Limited life (governed by protein stability) of a biocatalysist is seen as a stumbling block to the widespread exploitation of proteins in industry. Proteins are sensitive to heat or to other factors which may denature the molecule. Consequently, improving the heat stability of proteins has been one of the prime targets for protein engineering.

Argos and colleagues (1979) have devised a list of 'top ten' amino acid changes in a protein that could lead to increased thermostability. Of these, the change found most often was the replacement of lysine with arginine; the second most common change was the replacement of serine or glycine with alanine. More recent work with enzymes from thermophilic organisms (i.e. those that thrive in high-temperature habitats such as hot spings) including *Thermoplasma acidhilum* and *Pyrococcus furiosus*, has revealed many clues about how evolution has engineered stability into proteins (Rees and Adams, 1995; Russell *et al.*, 1994; Yip *et al.* 1995; Perutz, 1978). In thermophiles, the hydrophobic core of proteins is made more compact with a greater degree of hydrogen bonding (Rees and Adams, 1995). These interactions, in turn, confer rigidity to the structure which allows the enzymes to better resist the unfolding forces of thermal energy. Evidence from these studies indicates that compactness and less flexible regions of both helices and loops are important for thermostability but that if extreme resistance to heat is needed, strong electrostatic networks are also required. Yet, when comparisons of the overall structural folds in thermophiles and mesophiles are made, it is clear that thermostability is achieved with the minimum of alterations to the basic structural motifs.

The concept of preventing thermal motion by increasing structural rigidity has led to several attempts to modify disulphide bridges in proteins. As the majority of intracellular proteins do not have disulphide bridges, whereas a large number of secreted proteins do have these bridges, it seemed reasonable to assume that there must be a stability gain to having disulphide linkages. However, it has proved difficult to use protein engineering to provide the environment in a protein that is required for correct formation of a disulphide bridge, as discussed in more detail by Betz (1993).

In addition to improving the thermal stability of enzymes, protein engineering can also be used to change and improve enzyme activity and specificity. This has already been achieved in a number of enzymes of fundamental interest. For example, the catalytic mechanism of lactate dehydrogenase from *B. stearothermophilus* was altered by replacing arginine at position 109 with glutamine in order to retain bulk and hydrophobicity but remove the positive charge. This substitution triggered loop closure and thus exploited the favourable hydrophobic contacts made between the loop and the body of the protein. The enzyme became more active against malate thus becoming much more like malate dehydrogenase (Clarke *et al.*, 1986). Another example of specificity being altered was the conversion of allosteric inhibition to activation in phosphofructokinase by mutation of glutamic acid 187 to alanine 187 (Lau and Fersht, 1987).

Probably the most dramatic change brought about by the techniques of protein engineering has been the enhancement of activity of the antioxidant enzyme superoxide dismutase. Superoxide dismutase is an important agent in preventing oxidative damage in living organisms. Its action is to dismute the superoxide radical O_2 to molecular oxygen and hydrogen peroxide. Chemical reactions in this enzyme centre around a copper ion at the bottom of the active site tunnel. The enzyme has evolved to have one of the fastest rates of reaction of any known enzyme limited only by rates of diffusion. A reaction rate of 2×10^9 M^{-1} s^{-1} was further improved by exchanging one of the glutamic acids (implicated in electrostatic guidance of the substrate to the active site) for glutamine (Getzoff *et al.*, 1992). This work showed that electrostatically facilitated diffusion rates can be increased by design, provided that the detailed structural integrity of the active site electrostatic network is maintained.

Commercial enzymes whose activity has been improved by protein engineering have started to appear in the detergent industry. For example, at Novo Nordisk, scientists have removed a negative charge near the active site of lipase by substituting the aspartic acid residue in the 96 position with leucine, a neutral amino acid. In this way, the enzyme has been made less hydrophobic and contact between the protein and the negatively-charged fatty stains on laundry has been facilitated. The new lipase has been shown to have improved performance in detergents particularly at low wash temperatures (Gormsen, 1995). In the food industry, commercial enzymes engineered for improved activity or specificity (as opposed to thermal stability) have yet to make an impact.

6.4.3.1 Lysozyme. Hen egg white lysozyme is used in some hard cheeses to prevent the 'late gas blowing' defect caused by the spore-forming bacterium *Clostridium tyrobutyricum*. Since stability was not a real problem in avian lysozyme, efforts were made to improve the sporicidal activity of the enzyme. The size of the molecule and its overall electrostatic charge were thought to play important roles in the overall efficiency of the enzyme. Attempts to alter the size concentrated on a particular loop of protein that is very mobile and is connected to the rest of the protein by two disulphide bridges. When this mobile loop was removed, activity was retained but the enzyme was less stable than the native version. Further refinement of the enzyme could be expected to give a stable, smaller molecule, highly active against bacterial spores. This work has been patented and is the beginning of a targeted approach to engineering protein molecules for use in the food industry (Pickersgill *et al.*, 1994).

The ubiquitous phage T4 lysozyme was one of the first enzymes to have an engineered disulphide bridge inserted. This rendered the protein more stable (Perry and Wetzel, 1984). Additional disulphides were introduced into T4 lysozyme by Matsumura *et al.* (1988) and a new bridge between residues 9 and 164 significantly increased stability. Crystallographically, the 9–164 bridge still had the conformational flexibility of the wild type. A bridge at amino acid positions 21–142 also increased stability but decreased activity. Crystallographic evidence showed that the choice of the residue replaced could reduce destabilization to >2 kcal/mol, whereas entropic stabilization from adding a disulphide gave 3.8–5.4 kcal/mole stabilization (Pjuru *et al.*, 1990). In the case of disulphides between positions 90–122 and 127–154, the strain associated with the addition of the bridges outweighed the entropic gain. Matthews (1993) has mutated a polyalanine stretch into a T4 phage lysozyme helix to obtain increased thermal stability. The same group has conducted extensive mutagenesis and has stabilized both the electrostatic dipole of helices and also enhanced the helix forming potential (Nicholson *et al.*, 1988).

6.4.3.2 Subtilisin. Subtilisin, an enzyme used extensively in washing powders, is a protease that has a mechanism based on serine, histidine and aspartic acid. In the food industry, proteases are used to remove hazes from canned beers, tenderize meat, remove collagen in gelatin manufacture and prepare hydrolysed proteins for food flavouring. Much research has been directed towards improving the stability of proteases. In its original form, subtilisin was very easily oxidized. This instability was successfully improved upon by the removal of a methionine and substitution with a non-oxidizable amino acid.

Like the T4 phage lysozyme described above, subtilisin does not naturally have disulphide bridges, but unlike lysozyme, it also lacks free cysteines. Pantoliano *et al.* (1987) analysed the main chain configuration of subtilisin so that atoms in an ideal position for insertion of disulphide bridges could be established by reference to a database of natural disulphide bridges. A bridge between residues 22 and 87 was constructed and was reported to increase enzyme stability by 3.1°C. However, Mitchinson and Wells (1989) made five other disulphide bridges (not including 22–87) and found that none of them rendered the enzyme more stable than the wild type. In an earlier paper, Wells and Powers (1986) had made the 22–87 disulphide bridge in subtilisin and found no net stabilization. Pantoliano *et al.* (1987) pointed out that the Wells and Powers mutation also contained tyrosine 21 converted to alanine, but there still remains a discrepancy in the results.

Mitchinson and Wells (1989) used the programme PROTEUS to decide where to insert disulphide bridges in subtilisin. Comparisons to a related protease (protease K) which does have disulphide bridges were also used before a final decision was made. Disulphide bond strength was a poor predictor of stabilization to irreversible denaturation. However, in some cases disulphides were introduced within a strong calcium binding site.

At the time of writing, the introduction of disulphide bridges into subtilisin has not universally resulted in increased thermostability. It is important to remember two facts about disulphide bridges: (1) the bridges do not occur in all proteins, in fact they are rare in intracellular proteins; and (2) unfolded proteins often still have disulphide bridges. Thus, unlike hydrophobic forces and hydrogen bonds, disulphide bridges are not essential for correct folding of a protein and consequently for biological activity. Subtilisin has a folding pathway that works very well without disulphides and so, optimizing the stability by introducing a new bond to a molecule must be allied to an understanding of the free energy of the unfolded state.

6.4.3.3 Glucose isomerase. The importance of the high fructose corn syrup industry has already been discussed in section 6.3.1. In an effort to improve efficiency, enzyme manufacturers are beginning to use the powerful tools of protein engineering to change the temperature and pH stability of the enzymes used in the HFCS process. The conditions under which native glucose isomerase is used in the industrial scale process are arrived at as a result of a compromise between the optimum reaction temperature for the isomerization and the rate of enzyme denaturation. Using site-directed mutagenesis, scientists have recently been able to improve the stability of glucose isomerase by replacing some of the lysine residues in the protein with arginine.

Glucose isomerase is actually a D-xylose isomerase (EC 5.3.1.5) catalysing the isomerization of the five carbon sugar D-xylose to the ketose D-xylulose (Figure 6.4). It is also able to convert D-glucose (a six carbon analogue of xylose) to D-fructose. The enzyme is widely known as glucose isomerase even though its kinetic parameters are less favourable for the six carbon than for the five carbon substrate. The isomerization requires divalent metal cations such as Mg^{2+}, Co^{2+}, or Mn^{2+}. Bacterial xylose isomerases have four identical subunits each with a molecular mass near 43000. The monomer has a main domain which has the TIM barrel motif with a C-terminal helical domain which forms a large loop embracing a neighbouring subunit. The active site is located near the centre of the barrel, as found in other enzymes of this type. The first metal site is bound by four carboxylates of glutamic acid 181, 217, and asparagine 245 and 292. These residues are located on β-strands 5, 7 and 8. Each carboxylate donates an oxygen atom to the metal. They are at a distance of 2.0 Å which implies direct coordination of the metal. The second metal binding site is located less than 5 Å from the first site, nearer the surface of the TIM barrel. Residues involved in the second site belong to β-strands 6 and 7 and to the loops connecting these β-strands to the next alpha-helices in the barrel. The carboxylates of glutamic acid 217, asparagine 255 and asparagine 257, the imidazole group of histidine 220 and a well defined water molecule (water 690) constitute this site. When the substrate binds, the second metal moves and some of the binding to the protein is lost. The movement is towards the substrate, away from asparagine 257. The bonds to this residue and asparagine 255 break; those to glutamic acid 217, histidine 220 and water 609 are maintained as the metal carries with it the side chain of histidine 220 and a water molecule. The cyclic alpha-anomer of the sugar binds to the dimetal-enzyme complex through its oxygen atoms to the primary metal cation. The ring of the bound sugar opens and the extended form remains bound to the cation.

The enzyme with cation present is stable at 80–85°C but in the presence of glucose, activity is rapidly lost due to nonenzymatic chemical modification of important amino groups in the protein molecule. Such reactions between glucose and the amino side chains of lysine are well documented.

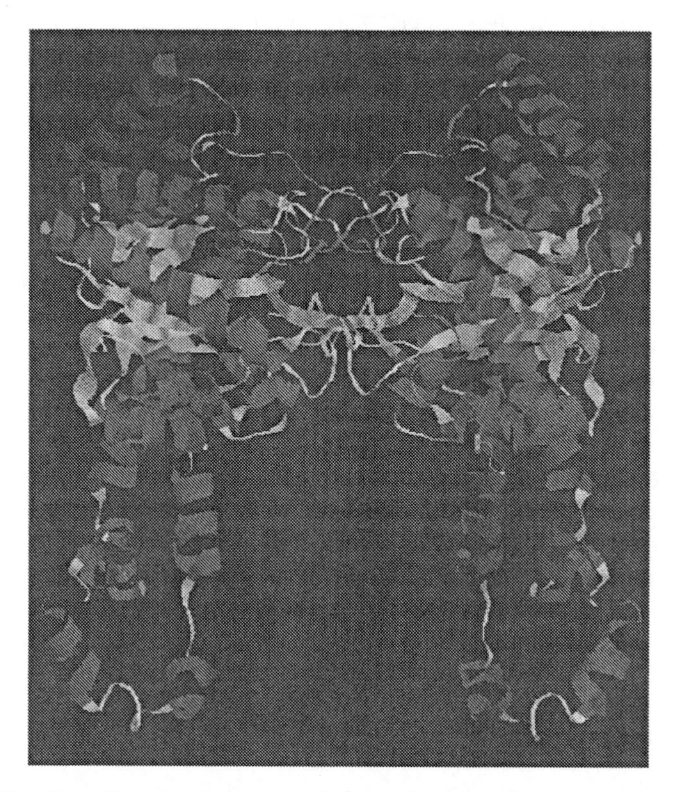

Fig. 6.4 The three-dimensional structure of glucose isomerase showing how the monomer associates to give a tetrameric enzyme. The main body of the enzyme is a triosephosphate isomerase (TIM) barrel structure and the amino acid which is mutated is in a loop joining the secondary structural elements.

The guanidinium group of arginine has reduced chemical reactivity due to a very high pKa value (12.4); the result of this is the maintenance of a positive charge in nearly all biological situations and there is resonance stability with an extended hydrogen-bonded network. In glucose isomerase, there are 20 lysines in each subunit and five of these have low solvent accessibility. The areas which have low solvent accessibility are buried in the tetramer and two of these lysines are found in the dimer interface (positions 253 and 100). Position 253 could accommodate an arginine side chain but not position 100.

Lysine residues at positions 309, 319 and 323 are located on the solvent-exposed side of helix 8 of the TIM barrel and form part of a cluster of positive charges with arginine 313, 321, 326. This close arrangement of lysine and arginines was thought to influence the pKa of the lysines and make them more reactive with glucose. The lysines at these positions were changed to arginine and the chemical reactivity of the protein in the presence of glucose was tested. The mutants produced by site-directed mutagenesis showed some increase in stability without loss in activity. In particular, a triple mutant 253 showed a six-fold gain in stability over the wild type enzyme at 60°C in the presence of glucose. The exact position of the mutant 253 lysine sidechain was shown by detailed X-ray analysis. The lysine lies at the end of a long loop which connects β- strand 7 to alpha-helix 7. This part of the loop was in van der Waals contact with its counterpart in a neighbouring subunit. Subunit packing left large cavities filled by water molecules which are ill-defined in the electron density map. The side chain of lysine 253 in subunit A extended into such a cavity and made contact with two other subunits involving asparagine 85, glycine 189 and aspartic acid 190 of subunit C and asparagine 225 of subunit D. There were therefore direct hydrogen bonds formed with some of these residues and the substrate (Mrabet *et al.*, 1992; Quax *et al.*, 1991).

The practical result of the engineering exercise has been the creation of a new glucose isomerase with a half-life of 1550 h at 60°C compared with 607 h for the wild-type enzyme (Quax *et al.*, 1991). Although the protein-engineered enzyme was thought to be in commercial use by the mid-1990s, the extent of its market penetration is not known due to commercial sensitivity of such information.

6.5 Future prospects

It has been predicted by sources from Novo Nordisk, the world's biggest manufacturer of enzymes, that almost all industrial enzymes will be produced from recombinant microorganisms by the year 2000 (Dalboge, 1995). Although others in the industry may question the timing, few would dispute the likelihood of recombinant organisms becoming the standard sources of many enzymes in the near future, primarily because of the numerous benefits that will accrue to both enzyme manufacturers (e.g. higher yields) and enzyme users (e.g. constancy of supply, improved stability, low price).

Many organisms currently used in enzyme manufacture produce mixtures of several enzyme activities. For example, some fungal pectinases used in fruit juice processing contain up to 20 different enzyme activities. Although such crude enzyme mixtures are likely to remain on the market because of their low cost and adequate performance in certain applications, it is conceivable that much more sophisticated, tailor-made enzyme mixtures prepared from a selected range of single-activity recombinant enzymes will become available in the future, particularly for applications where side-activities are undesirable.

Enzymes from recombinant organisms allow the replacement of their native counterparts without alteration of their structure and function. Protein engineering goes one step further than this by altering the structure of proteins to dramatically enhance the stability, efficiency and specificity of enzymes. Although still hampered by gaps in fundamental knowledge and the relatively high expense of R&D, protein engineering is set to become the most powerful tool available to generate new improved biocatalysts in the medium-term future.

References

ACNFP (Advisory Committee on Novel Foods and Processes) (1989–present) Annual Reports. Available from: ACNFP Secretariat, MAFF, Ergon House, Nobel House, 17 Smith Square, London SW1P 3JR.

AMFEP (The Association of Manufacturers of Fermentation Enzyme Products) (1996) Safety aspects of Enzymes Produced by Genetically Modified Micro-organisms. AMFEP Secretariat, Avenue de Corntenbergh 172, B-1040 Brussels, Belgium. 7 pp.

Anon. (1996) Company News. *Food Technology*, December, 31.

Anon. (1993) Rennet substitute affirmed as GRAS. *Food Technology*, July, 72.

Arbige, M. V. and Pitcher, W. H. (1989) Industrial enzymology: a look towards the future. *TIBTECH*, 7, 330–5.

Argos, P. *et al.* (1979) Thermal stability and protein structure. *Biochemistry*, 18, 5698–703.

Barbano, D. M. and Rasmussen, R. R. (1991) Cheese yield performance of fermentation-produced chymosin and other milk coagulants. *J. Dairy Science*, 75, 1–12.

van den Berg, G. (1991) Fermentation-produced chymosin; technological aspects of its use for cheesemaking. *IDF Cheese and Hygiene Week* (13–17 May 1991), Ede, The Netherlands.

van den Berg, J. A. *et al.* (1990) Kluyveromyces as a host for heterologous gene expression: expression and secretion of prochymosin. *Bio/technology* 8, 135–9.

Betz, S. F. (1993) Disulphide bonds and the stability of globular proteins. *Protein Science*, 2, 1551–8.

Bines, V. E., Young, P. and Law, B. A. (1989) Comparison of Cheddar cheese made with a recombinant calf chymosin and with standard calf rennet. *Journal of Dairy Research*, 56, 657–67.

Broome, M. C. and Hickey, M. W. (1990) Comparison of fermentation produced chymosin and calf rennet in Cheddar cheese. *Australian Journal of Dairy Technology*, 45, 53–67.

CCFRA (Campden and Chorleywood Food Research Association) (1997) *Biotechnology Club Bulletin*. Issue number 4. 15 pp.

Chothia, C. (1988) The 14th barrel rolls out. *Nature*, 333, 598–9.

Chothia, C. (1992) One thousand families for the molecular biologist. *Nature*, 357, 543–4.

Clarke, A. R. *et al.* (1986) Site-directed mutagenesis reveals role of mobile arginine residue in lactate dehydrogenase catalysis. *Nature*, 324, 699–702.

Committee on the Toxicity of Chemicals in Food (1992) Guidelines for the Safety Assessment of Microbial Enzyme Preparations in Food. Available from: Department of Health, Skipton House, 80 London Road, London SE1 6LW, UK.

CPC International, Inc. (1984) Affirmation of GRAS status of alpha-amylase enzyme from *Bacillus subtilis*. GRASP 4G0293. As referenced in: B. P. Sharma, 1989.

Curtiss, R. (1978) Biological containment and cloning vector transmissiblity. *Journal of Infectious Diseases*, 137, 668–79.

Dalboge, H. (1995) The impact of molecular biology on industrial enzymes now and in the future. *Proceedings of the European Congress of Biotechnology* 7, Abstract LUC 13.

Dornenburg, H. and Lang-Hincrichs, C. (1994) Genetic engineering in food biotechnology. *Chemistry and Industry*, 13, 506–10.

Emtage, J. S. *et al.* (1983) Synthesis of calf prochymosin (prorennin) in *Escherichia coli*. *Proceedings of the National Academy of Sciences*, 80, 3671–772.

Enzyme Biosystems, Inc. (1986) A petition for the affirmation of the GRAS status of amylase of *Bacillus megaterium* derived from *Bacillus subtilis*. GRASP 7G0328. As referenced in Sharma, 1989.

FDA 'Red Book' (1982) Toxicological Principles for the Safety Assessment of Direct Food Additives and Color Additives Used in Food. US FDA, Bureau of Foods.

FDA (1984) Lactase from *Kluyveromyces lactis*. GRAS affirmed, 21 Code of Federal Regulations (CFR) 184.1388. 49, *Federal Register* 47387, December 4.

FDA (1990) Direct food substances affirmed as generally recognized as safe: chymosin enzyme prepartion derived from *Escherichia coli* K12. *Federal Register* 55 (57), 10932.

FDA (1992) Chymosin from *Kluyveromyces lactis*. GRAS affirmed, 21 Code of Federal Regulations (CFR) 184.1685. 57, *Federal Register* 6476–9, February 25.

Food Chemicals Codex (1981) National Academy of Science/National Research Council, Food and Nutrition Board, Committee on Codex Specifications, National Academy Press, Washington DC, pp. 107–10.

Fordham, J. R. and Block, N. H. (1987) Regulatory issues of enzyme technology. *Developments in Industrial Microbiology*, 28, 25–31.

Getzoff, E. D. *et al.* (1992) Faster superoxide dismutase mutants designed by enhancing electrostatic guidance. *Nature*, 358, 347–51.

Gilson, M. K. and Honig, B. H. (1987) Calculation of electrostatic potentials in an enzyme active site. *Nature*, 330, 84–6.

Godfrey, T. and West, S.I. (1996) Introduction to industrial enzymology. In: Industrial Enzymology, 2nd edn. Eds T. Godfrey and S. I. West. Macmillan Press Ltd., London. pp. 1–8.

Goffeau, A., Oliver, S. and Dujon, B. (1996) Final conference of the yeast genome sequencing network, Trieste, Italy. Yeast ferments genome-wide functional analysis. *Nature Genetics*, 14, 1–2.

Goodenough, P. W. (1995a) A review of protein engineering for the food industry. *Int. J. Fd. Sci. Technol.*, 30, 119–39.

Goodenough, P. W. (1995b) Food enzymes and the new technology. In: Enzymes in Food Processing, 2nd Edn. Eds G. A. Tucker and L. F. J. Woods. Chapman & Hall, London. pp. 41–113.

Gorbach, S.L. (1978) Recombinant DNA: An infectious disease perspective. *Journal of Infectious Diseases*, 137, 615–25.

Gormsen, E. (1995) New Lipolase Ultra binds better to fat. *BioTimes*, 3, 12–13.

Green, M. L. *et al.* (1985) Cheddar cheesemaking with recombinant calf chymosin synthesised in *Escherichia coli*. *Journal of Dairy Research* 52, 281–6.

Guinee, T. P. and Wilkinson, M. G. (1992) Rennet coagulation and coagulants in cheese manufacture. *Journal of the Society for Dairy Technolology*, 45, 94–104.

Gustchina, E. *et al.* (1996) Post X-ray crystallography studies of chymosin: the existence of two structural forms and the regulation of activity by the interaction with the histidine-proline cluster of k-casein. *FEBS Letters*, 379, 60–2.

Hamer, R. J. (1995) Enzymes in the baking industry. In: *Enzymes in Food Processing*. Eds G. A. Tucker and L. F. J. Woods. Blackie Academic and Professional, Chapman & Hall, London. pp. 190–222.

Harris, T. J. R. *et al.* (1995) The structure of catalytic core of the family-F xylanase from *Pseudomonas fluorescens* and identification of the xylopentaose binding sites. *Structure* 2, 1107–16.

Harris, T. J. R. *et al.* (1982) Molecular cloning and nucleotide sequence of cDNA coding for calf preprochymosin. *Nucleic Acid Research*, 10, 2177–85.

Hicks, C. L., O'Leary, J. and Bucy, J. (1988) Use of recombinant chymosin in the manufacture of Cheddar and Colby cheese. *Journal of Dairy Science*, 71, 1127–35.

Hodgson, J. (1994) The changing bulk biocatalyst market. *Bio/Technology*, 12, 789–90.

JECFA (1984) General specifications for enzyme preparations used in food processing. FAO *Food and Nutrition Paper*, 31/2, 129–31.

JECFA (1987) *Principles for the Safety Assessment of Food Additives and Contaminants in Food*, World Health Organisation, Geneva, pp. 135–7.

Jenkins, J. *et al.* (1995) β-glucosidase, β-galactosidase, family A cellulase, family F xylanase and two barley glucanases form a superfamily of enzymes with 8-fold β/α architecture and with two conserved glutamates near the carboxy-terminus ends of β-strands four and seven. *FEBS Letters*, 362, 281–5.

Kawaguchi, Y. *et al.* (1987) Production of chymosin in *Escherichia coli* cells and its enzymatic properties. *Agricultural and Biological Chemistry*, 51, 1871–9.

Kumosinski, T. F. and Farrell, H. M. (1994) Solubility of proteins. Protein-salt water interactions, in *Protein Functionality in Food Systems*. Eds N. S. Hettiarachchy and G. R. Zieglar.
Lau, F. and Fersht, T. K. (1987) Conversion of allosteric inhibition to activation in phosphofructokinase by protein engineering. *Nature*, 326, 811–12.
Maat, J. *et al.* (1983). DNA molecules comprising the genes for preprochymosin and its maturation forms and microorganisms tranformed thereby. European Patent Application 077 109.
McCaman, M. T., Andrews, W. H. and Files, J. G. (1985) Enzymatic properties and processing of bovine prochymosin synthesized in *Escherichia coli*. *Journal of Biotechnology*, 2, 177–84.
MAFF Food Advisory Committee (1994) Annual Report. HMSO, London. 26 pp.
Marie-Henrietta, D., Gunther, H. and de Troostembergh, J. (1985) Process of enzymic conversion. European Patent Application No. 8530 3493.2. As referenced in B. P. Sharma, 1989.
Marston, F. A. O. *et al.* (1984) Purification of calf prochymosin (Prorennin) in *Escherichia coli*. *Bio/Technology*, 2, 800–5.
Matsumara, M. *et al.* (1989) Stabilization of phase T4 lysozyme by engineered disulphide levels. *Proceedings of the National Academy of Sciences*, 86, 6562–6.
Matthews, B. W. (1993) Structural and genetic analysis of protein folding and stability. *Current Opinion in Structural Biology*, 3, 589–93.
Mitchinson, C. and Wells, J.A. (1989) Protein engineering of disulphide bonds in subtilisin BPNI. *Biochemistry*, 28, 4807–15.
Mrabet, N. T. *et al.* (1992) Arginine residues as stabilizing elements in proteins. *Biochemistry*, 31, 2239–53.
Newland, J. W. and Neill, R. J. (1988) DNA Probes for Shiga-like toxins I and II and for toxin-converting bacteriophages. *Journal of Clinical Microbiology*, 26, 1292–317.
Nicholson, H., Becktel, W. J. and Matthews, D. W. (1988) Enhanced protein thermostability from designed mutations that interact with α-helix dipoles. *Nature*, 336, 651–6.
Olsen, H-S. (1995) Enzymes in Food Processing. In: *Biotechnology*, Vol. 9, *Enzymes, Biomass, Food and Feed*. Eds G. Reed and T. W. Nagodawithana. VCH Weinheim. 663–736.
Pantoliano, M. W. *et al.* (1987) Protein engineering of subtilisin BPNI: Enhanced stabilization through the introduction of two cysteines to form a disulphide bond. *Biochemistry*, 26, 207–2082.
Pariza, M. W. and Foster, E. M. (1983) Determining the safety of enzymes used in food processing. *Journal of Food Protection*, 46 (5), 453–68.
Pedersen, S., Jensen, B. F. and Jorgensen, S. T. (1995) Enzymes from genetically modified microorganisms. In: *Genetically Modified Foods. Safety Issues*. ACS Symposium Series 605. Eds K-H. Engel, G. R. Takeoka and R. Teranishi. American Chemical Society, Washington, DC, pp. 196–208.
Perry, L. J. and Wetzel, R. (1984) Disulphide bond engineered into T4 lysozyme: stabilization of the protein toward thermal inactivation. *Science*, 226, 555–7.
Perutz, M. F. (1978) Electrostatic effects in proteins. *Science*, 201, 1187–91.
Phillips, D. C. (1966) The three dimensional structure of an enzyme molecule. *Scientific American*, 215, 78–90.
Pickersgill, R. W. *et al.* (1994) The structure of *Bacillus subtilis* pectate lyase in complex with calcium. *Nature Structural Biology*, 1, 717–23.
Pickersgill, R. W., *et al.* (1994) Making a small enzyme smaller: removing the conserved loop structure of hen lysozyme. *Febs Letters* 347, 195–8.
Pjuru, P. E. *et al.* (1990) Structure of a thermostable disulfide bridge mutant of phage T4 lysozyme shows that an engineered cross-link in a flexible region does not increase the rigidity of the folded protein. *Biochemistry*, 29, 2592–8.
Praaning van Dalen, D. P. (1992) Application and regulatory postion of Maxiren. *Bulletin of the International Dairy Federation* (IDF) 269, 8–12.
Quax, W. J. *et al.* (1991) Stabilization of xylose isomerase by lysine residue engineeering. *Bio/Technology*, 9, 783–842.
Rees, D. C. and Adams, M. W. W. (1995) Hyperthermophiles – Taking the heat and loving it. *Science*, 3, 251–4.
Roller, S., Praaning-van Dalen, D. and Andreoli, P. (1994) The environmental implications of genetic engineering in the food industry. In: *Food Industry and the Environment*. Ed. J. M. Dalzell. Chapman & Hall, pp. 48–75.

Russell, R. J. M., *et al.* (1994) The crystal structure of citrate synthase from a hyperthermophilic Arachaeon, *Pyrococcus furiosus*. *Protein Engineering*, 8, 583–92.

Sambrook, J., Maniatis, T. and Fritsch, E.F. (1989) *Molecular Cloning: A Laboratory Manual*, 2nd edn. CSHLP, Cold Spring Harbor, NY, USA.

Schechter, I. and Berger, A. (1967) Mapping the active site of papain with the aid of peptide subsites and inhibitors. *Biochemical and Biophysical Research Communications*, 27, 157–62.

Scher, M. (1993a) Biotechnology's evolution spurs food revolution. *Food Processing*, January, 36–43.

Scher, M. (1993b) The human factor. *Food Processing*, April, 117.

Scientific Committee for Food (1991) *Guidelines for the presentation on food enzymes*. Reports of the Scientific Committee for Food, 27 series, EUR 14181, pp. 13–22.

Sharma, B.P. (1989) Genetic modification of enzymes used in food processing, in *Biotechnology and Food Quality*. Eds S. King, D.D. Bills and R. Quatrano. Butterworths, Boston. pp. 287–305.

Smith, D. D. S. *et al.* (1995) Greek key jellyroll protein motif design: Expression and characterisation of a first-generation molecule. *Protein Engineering* 8, 13–20.

Smith, H. W. (1978) Is it safe to use *Escherichia coli* K-12 in recombinant DNA experiments? *Journal of Infectious Diseases*, 137, 655–67.

Smith, R. A., Duncan, M. J. and Moir, D. T. (1985) Heterologous protein secretion from yeast. *Science*, 229 1219–24.

Sutcliffe, J. G. (1978) Nucleotide sequence of the ampicillin resistance gene of *E. coli* plasmid pBR322. *Proceedings of the National Academy of Sciences*, 75, 3737–46.

Tamuri, M., Kanno, M. and Ishii, Y. (1981) Heat and acid-stable alpha-amylase enzymes and processes for producing the same. US Patent 4,284,722.

Warwicker, J. and Watson, H.C. (1982) Calculation of the electric potential in the active site left due to α-helix dipoles. *Journal of Molecular Biology*, 236, 887–903.

Webb, E. D. (1992) *Enzyme Nomenclature*. Academic Press, San Diego. (Published for the International Union of Biochemistry and Molecular Biology.)

Wrotnowski, C. (1997) Unexpected niche applications for industrial enzymes drives market growth. *Genetic Engineering News*, February 1, 14, 30.

Yip, K. S. P. *et al.* (1995) The structure of *Pyrococcus furiosus* glutamate dehydrogenase reveals a key role for ion-pair networks in maintaining enzyme stability at extreme temperatures. *Structure*, 3, 1147–59.

Yoder, M. D., Keen, N. T. and Jurnak, F. (1993) New domain motif: the structure of pectate lyase C, a secreted plant virulence factor. *Science*, 260, 1503–7.

Zhang, X-J., Baase, W. A. and Matthews, B. W. (1991) Towards a simplification of the protein folding problem – a stabilizing polyalanine alpha-helix engineered in T4 lysozyme. *Biochemistry* 30, 2012–17.

Zoller, M. J. and Smith, J. (1982) Oligonucleotide directed mutagenesis using M13-derived vectors: an efficient and general procedure for the production of point mutations in any fragment of DNA. *Nucleic Acid Research*, 10, 6487–500.

7 Brewing with genetically modified amylolytic yeast

JOHN HAMMOND

7.1 Introduction

Yeast has a long history of industrial use, having been employed for the production of bread and alcoholic beverages since prehistoric times (Rose and Harrison, 1987). In all these processes yeast produces ethanol, carbon dioxide and numerous flavour compounds.

Originally, fermentations depended upon the wild yeast strains occurring naturally in the environment. During the 19th century, following the discovery of the microbial basis of fermentation (Pasteur, 1879) and the development of microbiology as a science, brewers began to isolate and produce pure yeast strains that guaranteed products of a more constant quality. For different beer types, different yeast strains were often employed. As new brewing processes were developed, new yeast strains were required and were isolated using entirely empirical methods. Large numbers of strains often had to be screened in order to obtain one with the desired property. With the advent of recombinant DNA (rDNA) methods, it has become possible to modify brewing yeasts in very specific ways to produce strains with desired new traits without affecting all the other properties which make the yeast suitable for beer production.

The first stage of beer manufacture is the production of wort which is subsequently fermented into beer by the addition of yeast. Wort is made by mixing ground malted barley with hot water and allowing the enzymes present to degrade the carbohydrates into a form which yeast can ferment. Because of the limited enzymic activity of malted barley and the limited fermentative ability of brewing yeast, about 20% of the carbohydrate extracted into wort remains in the beer. The majority of this carbohydrate consists of dextrins.

Low carbohydrate beers are often produced by the addition of commercial glucoamylase enzyme preparations to either worts, prior to boiling with hops, or to fermenting beer (Aschengreen, 1987). These enzymes digest the dextrins, releasing glucose which brewing yeasts can ferment. Commercial enzyme preparations, however, also have protein degrading activity which can have deleterious effects on beer quality, in particular reducing beer foam. One solution to this problem is to ferment conventional worts with a yeast having amylolytic properties. Such yeasts can not only be used to produce low carbohydrate beer, but have the added attraction of increasing the efficiency of conversion of cereal carbohydrate to alcohol during fermentation.

Attempts have been made over the years to produce amylolytic yeasts by conventional genetic means. These have been largely unsuccessful because classical hybridization techniques are ineffective with commercial strains of brewing yeasts. Attempts to overcome these difficulties by the use of so called 'rare mating' techniques and by spheroplast fusion methods (Hammond, 1996) have met with some success. However, although the strains generated produced extracellular glucoamylase enzymes, they also produced unpalatable beers because of the transfer of other genes during the relatively uncontrolled breeding experiments. The specificity inherent in recombinant DNA techniques has enabled these difficulties to be overcome.

7.2 The genetic modification procedure

Brewing yeasts are usually incapable of sporulation and so conventional breeding methods which depend upon spore formation cannot be used with these organisms. The genetic modification of laboratory strains of *Saccharomyces cerevisiae* is now a routine technique but the application of this procedure to commercial brewing yeasts presents a number of problems not usually encountered by the academic geneticist. First, brewing strains are polyploid in their genetic constitution and often do not contain a balanced number of chromosomes. Since such yeasts contain a number of copies of any specific gene, this means that marker genes inserted into brewing yeasts must confer a dominant character or they will not be detected. Secondly they are much more reticent to take up added DNA than the 'well-trained' laboratory strains. Nevertheless work has proceeded in many laboratories with brewing yeasts and there have been a number of successes.

7.2.1 *The genetics of amylolysis*

The yeast *S. cerevisiae* var. *diastaticus* produces an extracellular glucoamylase, which is capable of hydrolysing starch. This enzyme is encoded by a family of genes variously referred to as *STA* or *DEX*. The genes coding for *STA1* and *STA3* have been cloned (Yamashita *et al.*, 1985a,b) as has that encoding *STA2* (Meaden *et al.*, 1985). Interestingly, the three *STA* genes show strong DNA sequence homology suggesting that originally they were probably all derived from a common ancestral gene (Yamashita *et al.*, 1985b; Pretorius *et al.*, 1986).

The food-approved fungus *Aspergillus niger* also produces an extracellular glucoamylase, which is capable of hydrolysing starch (Ono *et al.*, 1988). This is the glucoamylase enzyme commonly used in brewing to produce low carbohydrate beers. The enzyme from this organism is produced commercially and sold by a number of companies for use in the food and beverage industries. The advantage of the *A. niger* glucoamylase, AMG, is that it possesses debranching activity, unlike the yeast glucoamylase, and can thus utilize a wider range of sugar substrates. The gene encoding this enzyme has also been cloned (Yocum, 1986).

7.2.2 *Yeast transformation methods*

For yeasts, a number of transformation procedures are available as illustrated in Figure 7.1. With all methods, DNA is first fragmented using an endonuclease (restriction enzyme). The fragments are then annealed to small circular DNA molecules called plasmids which have been linearized, usually with the same restriction enzyme. Recombinant plasmids are then generated by 'stitching-up' the joins using a DNA ligase enzyme. It is these plasmids which are used to introduce new properties into the recipient yeast cells. There are several ways in which yeast cells can be persuaded to take up DNA molecules. One technique involves removing the cell walls using a

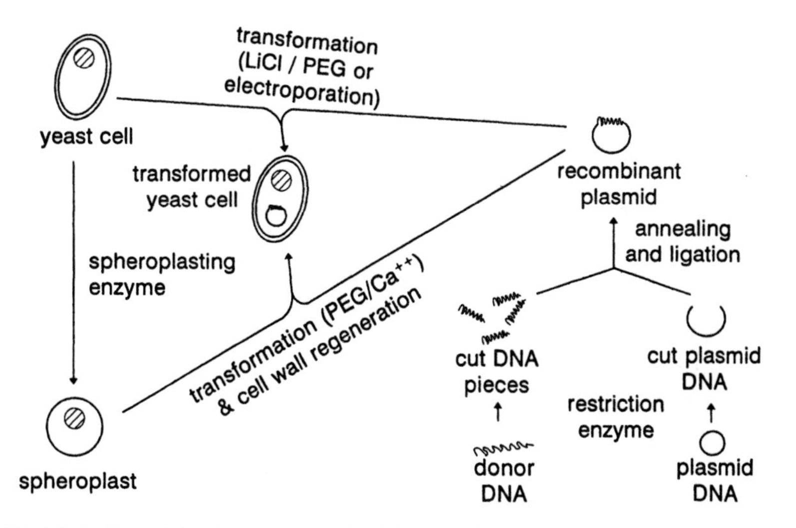

Fig. 7.1 Methods used for the transformation of yeasts. (Redrawn from Hammond, J. (1996) Yeast Genetics, in *Brewing Microbiology* (eds F. G. Priest and I. Campbell), Chapman & Hall, London.)

spheroplasting enzyme while another involves treatment of the yeasts with lithium salts. Alternatively DNA can be induced to enter yeast cells by the application of an electric current, a technique known as electroporation.

The new genes introduced into yeasts can be maintained in one of two ways, either they remain on the plasmids or they can be 'integrated' into one of the chromosomes of the yeast. Many yeast plasmid systems are based upon a naturally-occurring yeast plasmid called 2μm DNA. These so-called multi-copy plasmids have the advantage that, due to the number of copies present in the cell, there is usually good expression of the gene. However, plasmid-based systems are relatively unstable and the desired trait can gradually be lost with successive fermentations. In contrast, when the required gene is integrated into the yeast chromosome, any transformants are extremely stable, although the level of expression may be much reduced because of lower gene copy number, typically one copy per chromosome.

7.2.3 Construction of an amylolytic strain of brewing yeast using the STA2 *gene.*

The *STA2* gene cloned from *S. cerevisiae* var. *diastaticus* has been inserted into a recombinant plasmid pDVK2 (Vakeria and Hinchliffe, 1989). This plasmid does not possess DNA sequences derived from any organism other than *S. cerevisiae*. The detailed structure of pDVK2 is shown in Figure 7.2. It contains the following DNA elements:

1. 2μm DNA – a DNA fragment carrying the origin of replication of the yeast 2μm plasmid. This is essential for the maintenance of the recombinant plasmid within the transformed yeast cell. Without this, the plasmid pDVK2 would not replicate when yeast cells grow and divide and would be very rapidly lost from the population.
2. The *STA2* gene – this consists of the *STA2* structural gene cloned from *S. cerevisiae* var. *diastaticus* strain BRG 536 (Meaden *et al.*, 1985).
3. The regulation of expression of the *STA2* gene in pDVK2 is under the control of the *S. cerevisiae* phosphoglycerate kinase (*PGK*) gene promoter and terminator derived from plasmid pMA91 (Mellor *et al.*, 1983). These pieces of DNA ensure that the *STA2* gene is switched on so that the recombinant yeast produces glucoamylase enzyme at all times and under all conditions.
4. The *CUP1* gene of *S. cerevisiae* (Henderson *et al.*, 1985) is also carried by plasmid pDVK2. This gene is required as a selectable genetic marker for brewing yeast transformation since production of the glucoamylase cannot be used directly to select for yeasts which have taken up the recombinant plasmid during the transformation process.

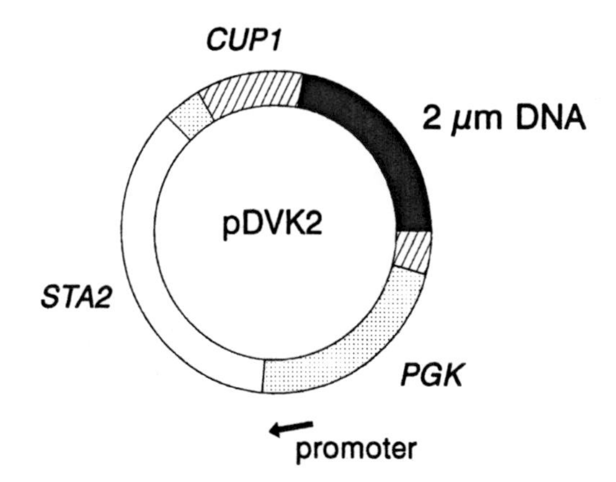

Fig. 7.2 Molecular map of plasmid pDVK2 derived entirely from strains of *S. cerevisiae*: 2μm yeast DNA; *STA2* gene from *S. cerevisiae* var. *diastaticus*; non-coding sequences derived from the *PGK* gene of *S. cerevisiae*, *CUP1* gene of *S. cerevisiae*.

As mentioned above, plasmid pDVK2 consists entirely of DNA sequences derived from the yeast *S. cerevisiae*. Consequently, pDVK2 is incapable of selection, replication and maintenance in the bacterium *Escherichia coli*. However, during the construction of pDVK2 it was necessary to pass DNA through an *E. coli* host. At this stage a modified version of pDVK2 was used which contained DNA sequences derived from the *E. coli* plasmid pBA112 (Andrews *et al.*, 1985). pBA112 is a derivative of the standard cloning vector pUC18 which carries the ColE1 plasmid origin of DNA replication and a functional *bla* gene encoding β-lactamase. The *bla* gene confers upon *E. coli* resistance to β-lactam antibiotics such as penicillin. However, once the procedures necessary to produce the functioning glucoamylase plasmid for yeast had been completed, the bacterial DNA sequences were completely removed to obtain plasmid pDVK2.

Plasmid pDVK2 was then used to transform lager brewing yeast strain NCYC1324 using the lithium salt procedure. The transformants were selected by virtue of their resistance to copper sulphate contained within the growth medium and were then screened for their ability to produce glucoamylase. All the copper-resistant transformants also produced glucoamylase as determined by their ability to hydrolyse maltotetrose.

One NCYC1324 [pDVK2] transformant was selected for further work and was checked for the presence of bacterial DNA sequences by genetic analysis and by testing for bacterial β-lactamase activity. All of these tests proved negative demonstrating the modified yeast to be free of bacterial DNA. NCYC1324 [pDVK2] was both copper-resistant and glucoamylase-positive (as evidenced by halo formation on starch plates and by enzyme assay of culture filtrates). These two properties were found to be co-inherited, as would be expected if they were both borne on the same plasmid. This was confirmed by independently measuring the relative copy number of the plasmid-borne *CUP1* and *STA2* genes; they were found to be equal at about 25 copies per cell.

7.2.4 An alternative approach to the construction of an amylolytic strain of brewing yeast using the GA *gene*

Another glucoamylase gene has been cloned from *Aspergillus niger* ATCC14916 (Yocum, 1986), a strain used for producing commercial glucoamylase for application in the food, brewing, and distilling industries. The donor *A. niger* strain is a mutant that secretes high levels of glucoamylase which has been used to produce food grade enzyme for many years. In particular, the enzyme from the donor organism has been used to produce light or low calorie beer in the United States since the 1970s.

The cloned gene has been integrated directly into the chromosomes of a commercial lager brewing yeast NCYC1342 to produce a second amylolytic brewing yeast strain. This strain contains an expression cassette in which the cloned gene specifying the *A. niger* glucoamylase is fused to the coding sequence for the yeast *STA4* signal peptide, the entire reading-frame being flanked by the high-level expression *ENO2* promoter and the *TPI1* terminator. These pieces of yeast DNA ensure that the glucoamylase will be efficiently produced and secreted from yeast. The expression cassette has been inserted into a specific yeast chromosomal site, the *HO* gene on chromosome IV. *HO* is responsible for mating-type switching in laboratory strains, but is redundant in a non-mating brewing strain. This means that disruption or deletion of this gene should not affect the brewing characteristics of NCYC1342. All of the copies of *HO* (of which there appear to be three) were disrupted in the recombinant strain NCYC1342.GA3. The portions of the donor DNA that were required for integration and expression of the glucoamylase gene were all derived from common laboratory strains of *S. cerevisiae*.

The procedure used to integrate the *GA* genes is shown in Figure 7.3. The host brewing yeast strain, NCYC1342, was transformed with the integrating plasmid pTW161, which had been cut to produce a piece of linear DNA. Transformants were selected on the basis of their resistance to the antibiotic geneticin (G418). Plasmid pTW161 contains the glucoamylase expression cassette flanked on both ends by yeast *HO* DNA. Outside the *HO* DNA, pTW161 contains the Tn903 G418 resistance gene driven by the yeast *CYC1* promoter. In addition it contains a bacterial *lacZ* gene which causes yeast cells containing it to grow as blue colonies on appropriate selective media.

In order to ensure complete stability the parent yeast was transformed three times in succession so that each of the three copies of the *HO* gene was disrupted. Plasmid pTW161 contains no yeast origin of replication, and so must integrate into a yeast chromosome to be maintained. After each

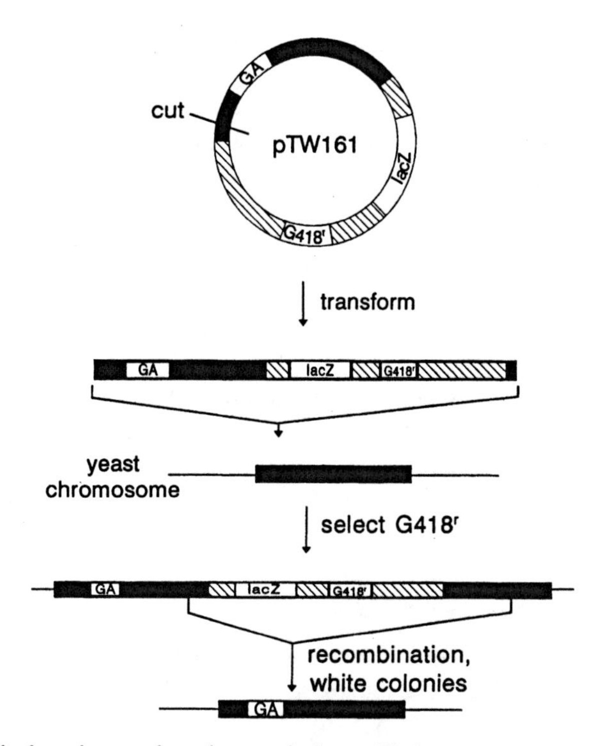

Fig. 7.3 Method used to produce the genetically modified yeast NCYC1342.GA3. Plasmid pTW161 was cut as indicated and used to transform yeast NCYC1342. Transformants were selected on the basis of their resistance to antibiotic G418 (Geneticin). Because of the presence of the *lacZ* gene, resistant colonies were also blue. Plasmid pTW161 contains the glucoamylase expression cassette flanked on both ends by yeast *HO* DNA. It contains no yeast origin of replication, and so must integrate into a yeast chromosome to be maintained. (Redrawn from Hammond, J. (1996) Yeast Genetics, in *Brewing Microbiology*, 2nd edn. (eds F. G. Priest and I. Campbell), Chapman & Hall, London.)

round of transformation, 'jettisoned integrants' which had lost the bacterial sequences were detected by loss of the blue coloration. Such recombinant strains were further screened for loss of antibiotic resistance and retention of glucoamylase secretion.

Each transformant that passed the screen was further analysed using standard laboratory genetic screening methods. A transformant that had the predicted structure at the *HO* locus, and no extraneous bacterial DNA was used for the next round of transformation. In this fashion, all three copies of the *HO* gene were occupied by the glucoamylase cassette, to give strain NCYC1342.GA3.

No plasmid vector sequences remained in the final strain NCYC1342.GA3, as confirmed by genetic analysis. Thus, the recombinant amylolytic yeast contains DNA elements which have been exclusively derived from *S. cerevisiae* and *A. niger*, both generally regarded as safe organisms.

7.3 The technological performance of genetically modified brewing yeasts

A prerequisite of most brewing yeast strain development programmes is that genetic modification should not adversely affect the intrinsic properties of the yeast strain. Indeed it is essential, when considering the use of gene technology to derive process improvements, that the fermentation and flavour characteristics of the beer resulting from the use of a genetically modified yeast are the same as achieved by the use of the unmodified parental yeast strain and current production methods. This requirement is a key feature of the work described in this chapter, since the objective of the genetic modifications was to facilitate the production of a particular quality of beer without recourse to the addition of exogenous enzyme. From a commercial point of view there would be little value in achieving this technical objective unless the beer produced by the genetically-modified yeast is similar in process and flavour characteristics to beer produced using the parental yeast.

7.3.1 The yeast strain NCYC1324 [pDVK2]

Yeast NCYC1324 [pDVK2] differs from its parent only by virtue of the presence of the recombinant plasmid pDVK2. In all other respects the two yeast strains are identical. Phenotypically the two yeast strains differ in that NCYC1324 is sensitive to copper, whereas NCYC1324 [pDVK2] is

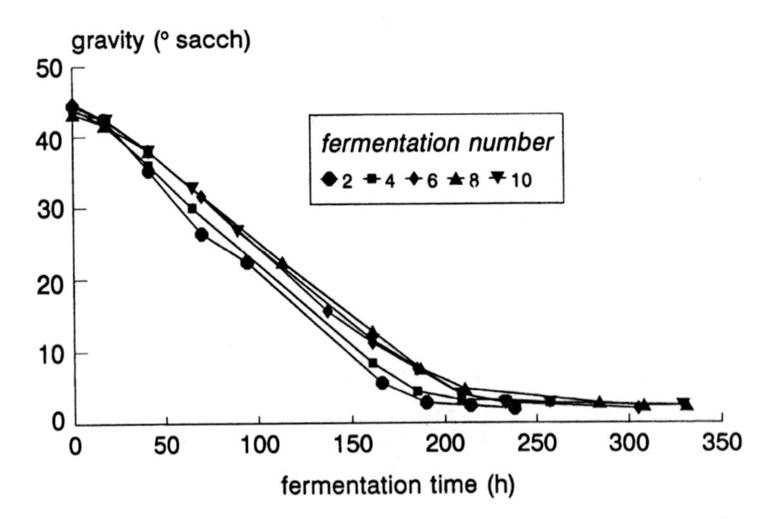

Fig. 7.4 Fermentation performance of NCYC1324 [pDVK2] in 10 consecutive pilot-scale fermentations. Fermentations were carried out at 12°C using 100 litres of lager wort. Yeasts were pitched at the rate of 5 g wet weight/litre into pre-aerated worts and fermentation followed by measuring the specific gravity of the wort. For clarity, data are presented from only the even-numbered fermentations.

resistant. In addition, the latter strain produces glucoamylase, the product of the *STA2* gene, which is secreted into the growth medium. This enzyme is a glucan 1,4 α-glucosidase which is capable of releasing glucose from α-1,4 linked chains, by sequential hydrolysis from the non-reducing ends of dextrin molecules. It is distinct from 'classical' glucoamylase in that it does not also hydrolyse α-1,6 glucosidic linkages and is therefore incapable of debranching dextrin molecules.

The known fermentation properties of the parental yeast strain NCYC1324 have been compared with those of the genetically-modified yeast at laboratory, pilot and commercial brewery scales. In addition, various laboratory tests have been performed to check that NCYC1324 [pDVK2] conforms to the behaviour expected of a brewing yeast.

At the laboratory scale, significant superattenuation was observed in the NCYC1324 [pDVK2] fermentations with no differences seen in the suspended cell counts between control and experimental fermenters. The sugar profiles showed a decrease in dextrin levels in the experimental beers. The fermentation performance observed in laboratory systems cannot necessarily be achieved in full-scale operations. This is particularly true of brewery fermentations where the generation of low levels of components responsible for the flavour of the final products is critical. Accordingly, it was important to investigate the performance of the genetically-modified yeast at the pilot scale and commercial scales.

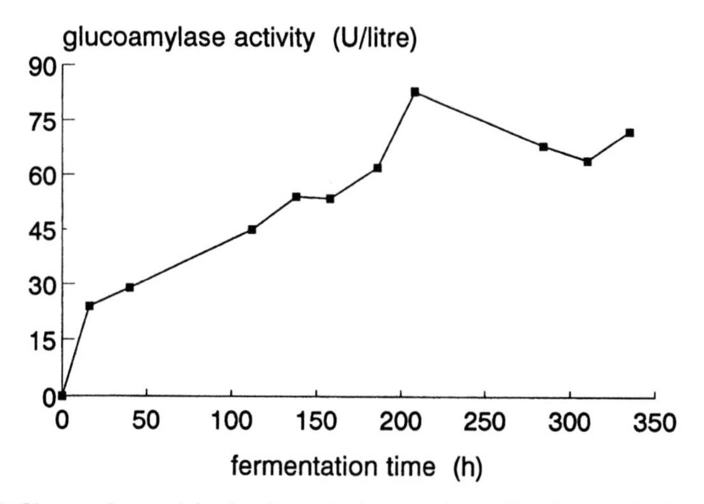

Fig. 7.5 Glucoamylase activity in pilot-scale fermentations using the genetically modified yeast NCYC1324 [pDVK2]. Samples of fermenting wort and beer taken from each of the fermentations shown in Figure 7.4 were analysed for glucoamylase activity. Enzyme activity is expressed as grams of glucose released per litre per hour from maltoheptose at 20°C. Glucose release was measured by HPLC. The data shown are for the eighth fermentation using NCYC1324 [pDVK2]. No enzyme activity was detected in the samples taken from fermentations conducted with NCYC1324.

Ten consecutive fermentations were performed at the 100-litre scale using a constant temperature of 12°C. Control fermentations were also carried out with the parent yeast. All 10 fermentations with the genetically-modified yeast gave very similar results with respect to superattenuation (Figure 7.4). The final gravity achieved by the modified yeast was significantly lower than that attained by strain NCYC1324.

During fermentations with the transformed yeast a continuous increase in secreted glucoamylase activity was observed (Figure 7.5). This is of course different from the situation when exogenous enzyme is added to a fermentation, where glucoamylase activity is present at a high level throughout the fermentation. In practice, the activity secreted by the genetically-modified yeast is adequate to achieve the desired degree of superattenuation. The levels of enzyme activity measured towards the end of fermentation are comparable to the levels recommended by enzyme manufacturers for dosing into beer fermentations to achieve the same outcome.

Analyses of the volatile components in the final beers produced in each of the successive fermentations revealed considerable similarities (Table 7.1). Also, importantly, the volatile flavour components produced during fermentation by the two yeasts were such that the beers were indistinguishable from each other. Tasting data also showed good consistency from fermentation to fermentation. These results were sufficiently encouraging for production trials to go ahead in a small commercial brewery.

The fermentations were carried out in 100 hl capacity cylindroconical vessels in a small commercial brewing plant using normal production lager worts. Fermentation temperature, following free-rise from 12°C, was maintained at 16°C until consecutive specific gravity measurements were the same. Cooling was then applied to bring the temperature down to 3°C. Yeast was next cropped and the beer temperature reduced to 0°C and held for 7 days. The beers were finally filtered, bottled and pasteurized as normal commercial products.

Table 7.1 Volatile components of beers produced in pilot scale fermentations by the genetically modified brewer's yeast NCYC1324 [pDVK2] and its parent NCYC1324. The beers produced from one of the fermentations shown in Figure 7.4 were analysed for volatile components in the head-space by gas-liquid chromatography

Flavour volatile (mg/l)	NCYC1324 (parent strain)	NCYC1324 [pDVK2] (genetically modified strain)
acetaldehyde	2.9	3.3
ethyl acetate	10.6	14.4
iso-butyl acetate	0.03	0.04
n-propanol	3.6	4.2
iso-butanol	6.2	4.2
iso-amyl acetate	1.0	1.1
iso-amyl alcohol	34.3	32.4
ethyl hexanoate	0.12	0.14

With the parental strain NCYC1324, the gravity profiles (Figure 7.6) show that wort was attenuated to 1006.6. Using the genetically-modified strain NCYC1324 [pDVK2], the specific gravity profile reached 1003.4 by the end of fermentation. The additional fermentability was due to the breakdown of wort dextrins into fermentable glucose by the genetically-modified yeast (Figure 7.7). The beers produced were once again of excellent quality and very similar to those produced by the conventional yeast.

In general all 'recombinant' 2μm plasmids are inheritably unstable in *S. cerevisiae*. Plasmid pDVK2 is no exception; thus NCYC1324 [pDVK2] loses the plasmid at a low but measurable rate during prolonged periods of cell growth. The data presented in Figure 7.8 indicate that about 20% of the plasmids were lost during 10 successive fermentations. Assuming a typical value of three generations per fermentation, the rate of plasmid loss is approximately 0.5% per generation. It should be noted that copper resistance always segregates with amylolytic activity; thus cells which become copper sensitive are also incapable of producing extracelluar glucoamylase. As a result of this plasmid segregation, copper sensitive derivatives of NCYC1324 [pDVK2] are identical to NCYC1324; in other words the loss of plasmid pDVK2 results in reversion to the parental yeast type; no new strains are produced.

Conventional brewing practice demands that yeast undergo up to 15 sequential brewery fermentations, in which a proportion of the yeast cropped from one fermentation is used to inoculate the next. It is apparent

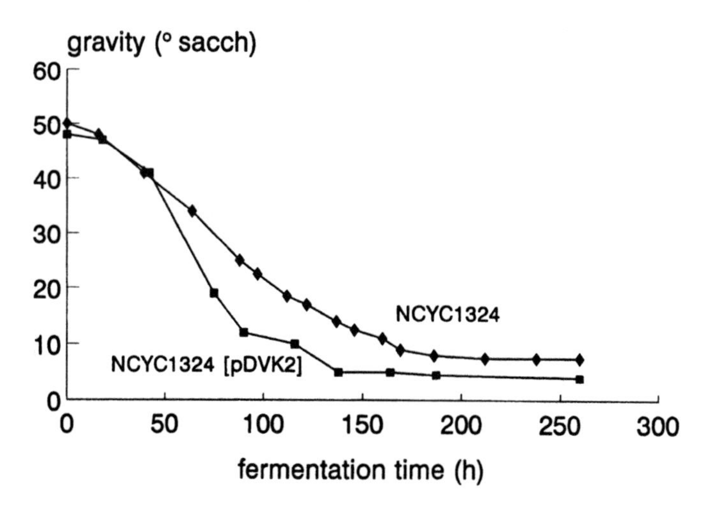

Fig. 7.6 Fermentation performance of NCYC1324 and its genetically modified derivative NCYC1324 [pDVK2] in a small commercial brewery. Fermentations were carried out at 16°C using 100 hl of commercial lager wort (Original Gravity 1050). Yeasts were pitched into pre-aerated worts and fermentation followed by measuring the specific gravity of the fermenting wort.

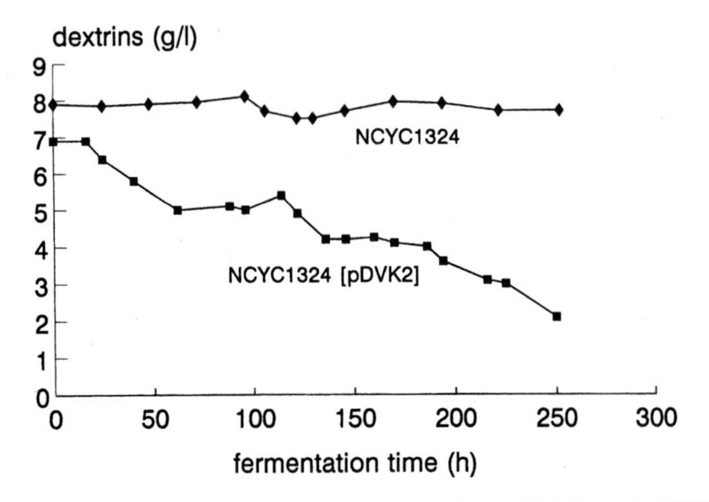

Fig. 7.7 Dextrin utilization by NCYC1324 and its genetically modified derivative NCYC1324 [pDVK2] during fermentation in a small commercial brewery. Samples of fermenting wort and beer taken from the fermentation shown in Figure 7.6 were analysed for dextrin content by HPLC.

from the data presented in Figure 7.8 that NCYC1324 [pDVK2] is sufficiently stable to complete satisfactorily a series of such full-scale commercial fermentations.

7.3.2 The yeast strain NCYC1342.GA3

The only difference between amylolytic yeast NCYC1342.GA3 and NCYC1342 is the presence of the glucoamylase expression cassette inserted into all three copies of the *HO* locus in the former strain. Phenotypically the two yeast strains differ only in that NCYC1342.GA3 produces an extracellular glucoamylase whereas the parental strain does not. Unlike the yeast glucoamylase described above this enzyme is a glucan 1,4 α-1,6 α-glucosidase which is capable both of releasing glucose from α-1,4 linked chains by sequential hydrolysis from the non-reducing end of the dextrin molecule and also of hydrolysing α-1,6 glucosidic linkages and hence debranching dextrin molecules.

The known fermentation properties of the parental yeast strain NCYC1342 have been compared with those of NCYC1342.GA3 at laboratory, pilot and commercial brewery scales. In addition, laboratory tests have been carried out to ensure that NCYC1342.GA3 conforms to the behaviour expected of a brewing yeast. Since the enzyme produced by NCYC1342.GA3 is commercially available it has also been possible to compare beers made with the parental yeast in the presence of the enzyme with those made by the genetically-modified yeast.

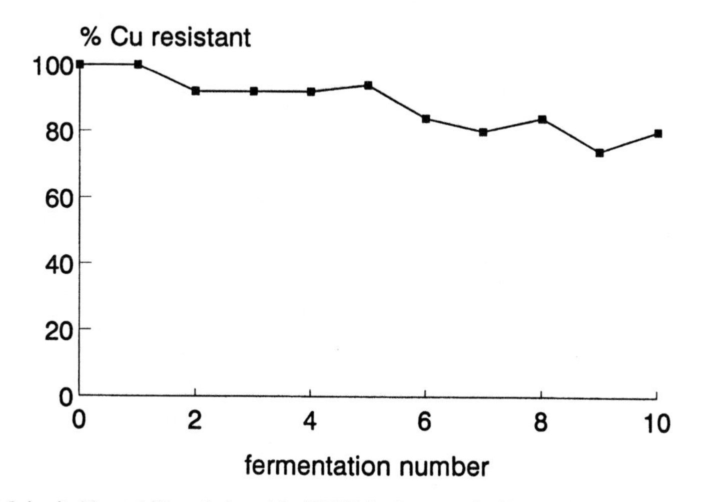

Fig. 7.8 Inheritable stability of plasmid pDVK2 in the genetically modified yeast NCYC1324 [pDVK2] during successive fermentations. A series of 10 sequential fermentations were carried out at 12°C in which yeast harvested from one fermentation was used to pitch the next fermentation. At the end of each fermentation, yeast samples were plated and individual colonies assessed for their copper resistance.

As for NCYC1324 [pDVK2], significant superattenuation and glucoamylase production was observed with NCYC1342.GA3 in laboratory fermentations and so detailed trials were carried out on the pilot and commercial scales.

NCYC 1342.GA3 was compared with NCYC1342, fermenting in the presence of commercial glucoamylase enzyme, in both 100-litre and 8-hl pilot fermenters. In both cases, the behaviour of the yeast in a number of successive fermentations was monitored, using the yeast cropped from one trial to repitch into the next. The recombinant yeast strain NCYC1342.GA3 was able to superattenuate a commercial wort, without the addition of enzyme preparations. In all trials, the fermentation profiles and sedimentation characteristics of the modified yeast and the normal yeast plus enzyme were very similar.

Beers produced from these trials were analysed by gas chromatography and the results showed a very similar pattern of volatile compounds for both beer types. Sugar analysis indicated a measurably higher level of residual glucose in the 'special beers' (NCYC1342.GA3) than in the normal beers (yeast plus enzyme) although this was not detected by tasting. The flavour profile of the beers produced using NCYC1342.GA3 compared favourably with the conventionally-produced products.

In fermentation trials at the 100-hl scale, the genetically-modified yeast fermented faster than the parental strain, at a rate which was very similar to that obtained with the parental yeast with added enzyme. The final gravity achieved by the modified yeast was significantly lower than that

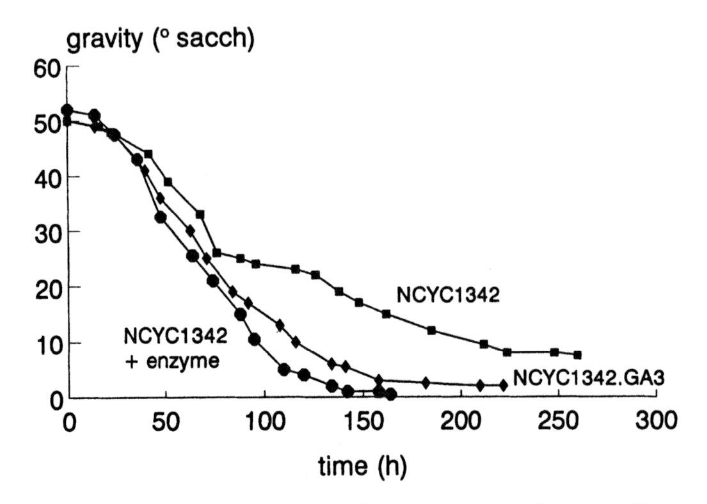

Fig. 7.9 Fermentation performance of NCYC1342 and its genetically modified derivative NCYC1342.GA3 in a small commercial brewery. Fermentations were carried out at 16°C using 100 hl of commercial lager wort (Original Gravity 1050). Yeasts were pitched into pre-aerated worts and fermentation followed by measuring the specific gravity of the fermenting wort. In the case of the fermentation marked NCYC1342 + enzyme, commercial glucoamylase was added to the wort immediately prior to pitching.

attained by strain NCYC1342 (Figure 7.9), although not as low as that achieved with the parental strain plus added enzyme. Average values for final attenuation, after maturation and bottling, were NCYC1342: 1006.3, NCYC1342.GA3: 1000.5 (5.8° sacch. superattenuation) and NCYC1342 + enzyme: 997.8 (7.6° sacch. superattenuation). The additional fermentability was due to the breakdown of the wort dextrins into glucose by either the added glucoamylase or that produced by the genetically-modified strain (Figure 7.10). This represented hydrolysis of 78% of the wort dextrin in the case of the added enzyme and 61% dextrin hydrolysis by the genetically-modified strain.

The difference in superattenuation between the two sets of glucoamylase fermentations can be accounted for by the presence of additional contaminating carbohydrase activities in the commercial enzyme preparation and the fact that the added enzyme was present throughout the fermentation whereas the enzyme produced by NCYC1342.GA3 was pure and only built up in activity towards the end of fermentation. The levels of enzyme activity achieved during fermentation are comparable to the levels recommended by enzyme manufacturers for dosing into fermenting beer.

In all other respects the yeasts behaved very similarly to each other and, importantly, the volatile flavour components produced during fermentation were such that the beers were indistinguishable organoleptically from each other (Figure 7.11). The beer made with added commercial enzyme had a very poor foam quality. This was not observed in any beer from

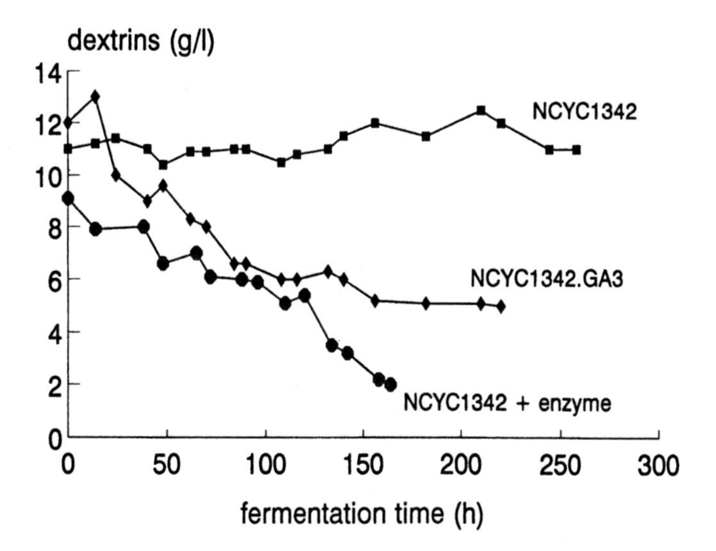

Fig. 7.10 Dextrin utilization by NCYC1342 and its genetically modified derivative NCYC1342.GA3 during fermentation in a small commercial brewery. Samples of fermenting wort and beer taken from the fermentation shown in Figure 7.9 were analysed for dextrin content by HPLC.

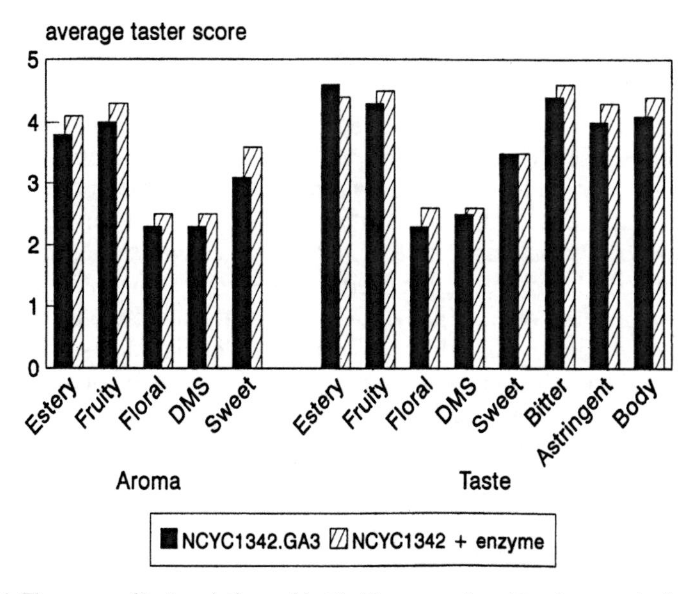

Fig. 7.11 Flavour profile descriptions of bottled beers produced by the genetically modified strain NCYC1342.GA3 and its parent NCYC1342 in the presence of commercial glucoamylase. Bottled, pasteurized beers from fermentations such as those shown in Figure 7.9 were tasted by an experienced flavour profile taste panel and the mean scores for the major descriptive flavour terms calculated.

fermentations using the genetically-modified strain or using the parental yeast without enzyme addition. This probably resulted from activity of contaminating proteases present in the crude enzyme preparation. Thus the beers produced by the genetically-modified amylolytic yeast strain were of excellent quality and in some respects superior to those produced by conventional procedures with added enzyme.

In general, it is expected that any gene integrated into the genome of *S. cerevisiae* will be no less stable than any of the genes normally present in the yeast. The transformation method employed was designed to maximize stability by ensuring that all copies of the *HO* gene contained an integrated glucoamylase gene. However since the procedure used was also designed to eliminate all unnecessary DNA from the yeast intended for commercial use, no selective resistance markers were retained within the yeast. Consequently it is not possible to employ any selective pressure during storage and use of NCYC1342.GA3 to ensure retention of the *A. niger* glucoamylase gene. However, the yeast has been used for numerous growth and fermentation experiments over a number of years without any evidence of loss of activity in the stock cultures. Additionally, the strain has been shown to be stable through a simulated 10^{12} litre scale up (70 generations). Clearly NCYC1342.GA3 is very stable and well able to complete satisfactorily a series of full-scale commercial fermentations.

7.3.4 *Advantages and disadvantages of genetically modified amylolytic yeasts*

For commercial use, an amylolytic yeast is preferable to using enzymes in a number of respects. Since the yeast produces a constant amount of enzyme during fermentation the brewer has more control over the degree of superattenuation achieved. Also, accidental over- or under-dosing of enzyme is avoided since a yeast-based system is, to some extent, self-correcting and the likelihood of major errors in dosing rate is higher when adding enzyme than when pitching yeast. Finally, and importantly, there are no side activities in the enzyme produced by the genetically-modified yeast. The major disadvantages are the need to obtain approval for use and consumer doubts about the use of genetic modification in the production of food and drink. These are discussed later in this chapter.

7.4 Health, safety and environmental issues

7.4.1 *Containment considerations*

The strains used for genetic modification (NCYC1324 and NCYC1342) are ‘standard’ strains of lager brewing yeast which have an extensive history of use for the commercial production of beer. They are also used in the pilot brewery of Brewing Research International for the production of lager beers. Consequently, there is a wealth of knowledge concerning the fermentation characteristics of the yeasts and the qualities of beers they produce. There is no evidence to suggest that these yeasts are capable of producing toxic metabolites; the long-standing use of these and similar brewing yeasts to produce commercial quantities of beer is in itself adequate evidence of this fact.

Nevertheless the yeasts are classified as genetically-modified and so all laboratory experiments undertaken during their development were performed at containment level 1 (Good Microbiological Practice) based on risk assessments carried out according to ACGM/HSE/DOE Note 7 (1993). All fermentation experiments were carried out either at containment level 1 (laboratory fermentations) or under conditions of good large-scale practice (pilot and commercial fermentations).

7.4.2 Transfer of 'novel' genetic material to other microorganisms and survivability of the modifed strains

The 'novel' genetic material present in NCYC1324 [pDVK2] is carried by the recombinant plasmid pDVK2. This plasmid is based upon the 2μm plasmid of *S. cerevisiae* carrying the 2μm origin of DNA replication; consequently pDVK2 is capable of replication and maintenance in all strains of *S. cerevisiae* which harbour the endogenous 2μm plasmid. In theory, this means that virtually all strains of *S. cerevisiae* are potential hosts for pDVK2. In practice, the 2μm circle is only transferred between yeasts during mating (Livingstone, 1977). Mating occurs when two haploid cells of opposite mating types come in contact; cell and nuclear fusion follow. Brewing yeasts usually lack a mating type and as a consequence are reticent to mate (Hammond, 1996). In this respect NCYC1324 [pDVK2] is no exception; thus, the transfer of pDVK2 to any other strain of yeast is likely to be an extremely rare event in nature. In any event the transfer of pDVK2 is unlikely to result in the prolonged survival of pDVK2 in the population due to the plasmid's innate inheritable instability (Figure 7.9). Two micron-based plasmids have not been shown capable of replicating in species other than *S. cerevisiae* nor have they been successfully transferred to other microorganisms. Consequently, it is very unlikely that pDVK2 would be transferred to microorganisms present in the human gut flora.

The 'novel' genetic material present in NCYC1342.GA3 is integrated into the *HO* locus of chromosome IV and so the loss of the gene or transfer to other organisms would be no more common than the transfer of other genes from the genome of *S. cerevisiae*. Transfer of genes between different strains of *S. cerevisiae* normally occurs during mating. As explained above, brewing yeasts such as NCYC1342.GA3 are reticent to mate; thus the mating and transfer of the *Aspergillus* glucoamylase genes to any other yeast is likely to be an extremely rare event. Any such mating would lead to strains that would be similar to the parent strain in the relevant properties. The modified strain NCYC1342.GA3 has a slight growth disadvantage compared with its parent, in most media, probably due to the burden of a relatively highly-expressed heterologous (i.e. not of the same species) secreted protein. Thus, the engineered strain or its descendants would not be expected to compete well in the environment except where starch is the predominant carbon source, such as in a rotting potato. Natural isolates of *S. cerevisiae* var. *diastaticus* that secrete active glucoamylase are well known, but these isolates are rather rare in the environment relative to those that don't secrete glucoamylase, suggesting that secretion of glucoamylase is not a highly selected trait for *S. cerevisiae*.

At the end of fermentation and conditioning, yeast is separated from beer by filtration. This process results in 'bright' beer which is substantially free from yeast. However, small numbers of yeast cells may traverse the filter and enter the final beer. Typically, bright beer will contain between 10–50 viable cells/ml. This beer will subsequently progress into bottle or can which is then tunnel pasteurized (typically 10–30 min. at 60°C). Tunnel pasteurization is very effective at eliminating all remaining viable yeasts from beer. Survival of live organisms into the final package is a very rare event. Any yeast remaining in the beer is consumed with the product and will therefore enter the human gut. Given the nature of the genetic modifications of NCYC1324 [pDVK2] and NCYC1342.GA3, these yeasts possess the same potential to survive in the human gut as their parental counterparts and available evidence strongly suggests that they will be rapidly eliminated in exactly the same way as normal strains of *S. cerevisiae* (Pecquet *et al.*, 1991).

7.4.3 Pathogenicity and allergenicity

The genetically modified brewer's yeasts are strains of the species *S. cerevisiae*. Yeast strains of this species are not pathogenic and there is no reason to suppose that NCYC1324 [pDVK2] and NCYC1342.GA3 would behave differently. Pathogenic yeasts belong to completely different genera from *Saccharomyces*. Their pathogenic character is determined largely by their cell-wall structure, which differs considerably from that of *S. cerevisiae* with respect to mannan structure, protein composition and antigenic structure.

It is generally agreed that the absence of pathogenic or toxic traits in a genetically modified organism is determined by the status of the host organism, unless genes coding for a pathogenic character are involved in the genetic transformation process. Thus, the modified brewer's yeast could not, in any conceivable fashion, exhibit any pathogenic character for the following reasons:

1. the host organism is not pathogenic;
2. the donor DNA has been derived entirely from strains of *S. cerevisiae* and *A. niger* which are not pathogenic;
3. in the modified strain, only yeast promoters and terminators are used;
4. the presence of glucoamylase activity and the conversion of dextrins to glucose are not in themselves harmful.

There have been no recorded instances of problems associated with the consumption of *S. cerevisiae*. This yeast is consumed by humans in bread and various fermented products, especially so-called 'traditional' cask beers and bottle-conditioned ales. Such beers contain high levels of viable yeast and have been drunk for centuries without problems associated with consumption of yeast. Moreover the number of live yeast cells normally ingested during the consumption of fresh fruits, which have a high level of surface contamination with *S. cerevisiae* and associated species, will be considerably greater than the levels of yeasts associated with the beers described here. Similarly, the high content of vitamin B, characteristic of *Saccharomyces* yeasts will not be in any way affected by the genetic modification and so it is not anticipated that the nutritional qualities of the yeast or of the beer made by it will be altered.

One difference between conventional beers and beers made using the genetically modified yeasts described here, is that the latter contain one of two glucoamylase enzymes. One of these enzymes (that produced by NCYC1342.GA3) is already found in a number of commercially available beers produced using the *Aspergillus* enzyme. The other protein, derived from *S. cerevisiae* var. *diastaticus* is not normally found in beer. The possibility of this giving rise to allergic reactions in persons consuming the beer can be discounted for two reasons. A closely related protein coded for by the *SGA* gene is produced by brewing yeasts and is, therefore, present in beer. Similarly, the glucoamylase coded by the *STA2* gene is present in foodstuffs such as bread and yeast extracts and so is part of the human food chain. No allergic reactions have been reported in these cases, thus making the likelihood of adverse reactions to beer produced by NCYC1324 [pDVK2] extremely unlikely.

7.4.4 Disposal of waste yeast – environmental considerations

The genetically modified yeast strains are intended for exploitation in industrial brewing plants, without any precautions with regard to disposal and containment other than those normally practised for the process. In this regard it should be noted that conventional brewing fermentations are frequently exposed to the atmosphere; consequently, some exposure of process operators to yeast biomass does occur. However, given the nature of the genetic modifications undertaken, any risk associated with operator exposure to these yeasts will be comparable to that which already exists for exposure to the parental counterparts.

Because of the nature of the brewing process, it is inevitable that the environment surrounding breweries contains higher concentrations of yeast than other areas. However, this has never been known to cause any problems. In this regard, the nature of brewing yeasts is important. The strains used for brewing have been selected over the centuries for optimal performance in brewery fermenters. These strains are adapted to grow best under rich, non-competitive conditions and have lost their vigour for survival in outside environments when compared with the wild type yeasts found naturally on plants, fruit, etc. Indeed, industrial yeast strains are rarely found in the environment and there is no evidence that growth and/or survival of brewer's yeast occurs around beer production plants. The secretion of glucoamylase enzyme by the modified yeasts will not, in any way, enhance their growth and survival since the enzymes are extracellular and any sugar produced will be available to all the natural flora of the environment which will be better able to take advantage of this than the modified brewer's yeasts. In this context it is worth pointing out that *A. niger*, the source of one of the glucoamylases, is a common environmental organism.

7.5 Legislative and labelling position

When genetic modification methods were first developed serious concerns about the technology were raised. These included the possibilities of uncontrolled transfer of genes to other organisms outside the laboratory, colonization of the environment by genetically modified organisms and inadvertent production of new pathogenic organisms. With the passage of time it has become generally evident that these potential problems were much exaggerated. However, because of the expression of these worries, a plethora of legislation has been enacted to address such issues. The regulations, although differing in detail, are broadly similar in most countries (Chapter 4). They recognize the need for different facilities for work with disparate genetically modified organisms, the degree of containment necessary varying with the perceived risk. All genetic modification of brewing yeasts has always fallen into the lowest containment category which usually requires only the facilities found in any good microbiological laboratory.

For a detailed discussion of the regulatory issues concerning genetically modified foods, the reader is referred to Chapter 4. Briefly, in the UK, a number of advisory committees are involved in the approval process including the Advisory Committee on Novel Foods and Processes (ACNFP) who are concerned with approval for consumption, the Food Advisory Committee (FAC) who advise on labelling, the Advisory Committee on Releases to the Environment (ACRE) who are concerned with environmental impact, and the Advisory Committee on Genetic Modification (ACGM) who consider health and safety matters. Clearly, progressing from R&D brewing to obtaining permission for commercial use of a genetically modified brewing yeast represents a considerable step forward for any brewer to undertake. It is this step that Brewing Research International set out to take, on behalf of its members, using one of the amylolytic yeasts described above as a test case.

In the UK, the ACNFP has set out in some detail the information required in any application for approval of a novel food whose manufacture involves the use of genetically modified organisms. The applicant has to submit details of:

1. The likely level of consumption of the novel food together with any relevant data concerning its nutritional value when compared with other foods it might replace.
2. The history of the parent organism and in particular its current role in food production.
3. The genetic make-up of the new genetically modified organism.
4. Toxicological data on the new food. The committee will accept reasoned argument why this is not necessary in specific cases.
5. The genetic methods used to construct the genetically modified yeast must be described.
6. The properties of the new strain must be described together with details of its genetic stability and the levels of expression of the novel gene.
7. Finally, in cases where live genetically modified microorganisms are to be consumed, data on survival of the organism and the likelihood of transfer of genes into bacteria in the gut must be provided. (In the case of pasteurized beer this is not, of course, required.)

The appropriate data were assembled and an application submitted to the UK Ministry of Agriculture, Fisheries and Food for the approval for commercial beer production of yeast NCYC1324 [pDVK2] containing the *STA2* gene from *S. cerevisiae* var. *diastaticus*. The main arguments for being granted permission to use this yeast in commercial production were fourfold:

1. The yeast had been produced by self cloning, that is the new yeast contained only DNA derived from *S. cerevisiae*.
2. There was no unnecessary DNA present derived from other organisms.
3. The yeast actually performed the functions required of it and led to greater efficiency of conversion of brewing raw materials into alcohol.
4. The yeast was sufficiently stable for commercial use.

On 16 February 1994, permission was received from the Agriculture and Health Ministers of Great Britain who concluded that there were no food safety reasons why the yeast should not be used to produce foodstuffs. Brewing Research International is now regularly making batches of beer with the genetically modified yeast which have been distributed widely and drunk by brewers from all over the world, representatives of the media and members of the public. The product is called Nutfield Lyte and is the first beer in the world made from a genetically modified yeast to be approved for production and general consumption.

The current position concerning labelling of foods and beverages made using genetic modification techniques is discussed in detail in Chapter 4. Briefly, in the UK, the Food Advisory Committee (FAC) has proposed labelling guidelines in 1993 which stipulated that a genetically modified food should be labelled if it contains a copy gene originally derived from a human or an animal which is the subject of religious dietary restrictions; and if it is a plant or microbial material containing a copy gene originally derived from an animal. These rules would not apply if the inserted copy gene had been destroyed by processing and was not therefore present in the food. Clearly, the beer made with the modified amylolytic yeasts did not fall into any of the categories requiring labelling and so, the main label developed for the beer does not contain any specific information concerning genetic modification. However, most retailers felt that it was important to keep their customers fully informed of any changes to products on their shelves and so, many companies making or intending to make products using genetic modification have voluntarily decided to openly label them. Accordingly, although the beer made using amylolytic yeasts is not on commercial sale, we have attached a back label to the bottled product which explains the process whereby it was made and the advantages of doing so.

7.6 Consumer acceptance and marketing

So far very few foods made using genetic modification have appeared on the market, but these few have been generally well received in the countries in which they have been released. The beer made from amylolytic yeasts is not on commercial sale but has been produced in significant quantities and consumed by many people including members of the general public. For the most part, reaction has been neutral or positive. Reports in the media have been largely factual and supportive although there has been some criticism from the environmental lobby and from the German media.

Despite the evidence so far, there is a perceived consumer reticence to purchase and consume genetically modified foods and beverages. Because of this concern, no genetically modified beer has so far appeared on sale anywhere in the world and a marketing strategy has yet to be developed. Certainly, as far as the small quantities produced by BRF International are concerned, there has been no attempt to disguise the nature of the beer: in fact its origin has been loudly proclaimed and we cannot meet demand for the beer. Perhaps there is a lesson to be learned from this.

7.7 Future prospects

Although the amylolytic yeast NCYC1324 [pDVK2] is the only brewing strain so far approved for commercial use, a number of groups worldwide have been using genetic modification techniques to develop a range of other brewing yeasts with widely varying properties. In years to come, some of these will undoubtedly find their way into commercial production. Most of the modifications have been concerned with improving the efficiency with which brewing yeasts utilize carbohydrates or with improvements in the flavours of the beers produced. A few examples of these are given below. For a fuller discussion the reader is referred to a recent review article on the subject (Hammond, 1995).

7.7.1 Improved utilization of carbohydrates: yeasts secreting β-glucanases

Barley β-glucanase is heat labile and easily destroyed during the malting and brewing processes. As a consequence, barley β-glucans are sometimes found in beer, where they present filtration difficulties and can give rise to hazes. To overcome this problem, microbial β-glucanases are often added to mashes, and so there has been interest in transferring genes coding for β-glucanase into brewing yeasts.

Genes have been cloned from *Bacillus subtilis* (Cantwell and McConnell, 1983), from *Trichoderma reesei* (Knowles *et al.*, 1985) and from barley (Jackson *et al.* 1986). When the β-glucanase gene from *B. subtilis* was expressed in yeast significant levels of extracellular β-glucanase were obtained. The fermentation performance of the transformed yeasts in laboratory fermenters was identical to that of the parent brewing yeast, but the beers produced had lower β-glucan contents and their viscosities and filterabilities were much improved (Lancashire and Wilde, 1987). Similar results were obtained when the fungal enzyme was expressed in yeast (Enari *et al.*, 1987).

The β-glucanase from barley shows highest activity at a pH of 4.7 whereas the bacterial enzyme is most active at pH 6.7. The barley enzyme is thus better suited to the conditions found in beer fermentations. The gene for this enzyme has been inserted into a yeast expression system and used to transform yeasts (Jackson *et al.*, 1986). The transformed strains all produced extracellular β-glucanase. A similar transformant of a brewing yeast has been used for pilot scale beer production (Berghof and Stahl, 1991) and once again filtration rates were increased without affecting flavour and foam stability.

7.7.2 Yeasts with a reduced capacity to produce diacetyl

The removal of diacetyl (which gives rise to a 'buttery' flavour considered undesirable in beer) formed during fermentation, requires brewers to carry out a time-consuming and expensive flavour maturation process. Diacetyl is formed in beer by a chemical reaction, from α-acetolactate released by yeast during fermentation. It is subsequently removed by yeast activity. Various genetic approaches have been attempted to reduce or eliminate diacetyl formation. Acetolactate is formed as an intermediate during synthesis of amino acids by yeast. Since the conversion of α-acetolactate to diacetyl is the rate limiting step, the best approaches have involved either preventing α-acetolactate formation (reducing the activity of the synthase enzyme) or removing it rapidly before it is converted to diacetyl by increasing the activity of enzymes later in the amino-acid synthetic pathway or by introducing a completely new enzyme activity (Figure 7.12).

Partial reduction in acetolactate synthase activity can be achieved by using mutants of the *ILV2* gene resistant to sulphometuron methyl, which inhibits the activity of this enzyme. Some resistant forms of the enzyme are less active and in this way strains of brewing yeasts have been produced which possess very low acetolactate synthase activities. Mating of these yeasts has yielded hybrids that ferment well but produce little diacetyl (Kielland-Brandt *et al.*, 1989).

An alternative approach to reducing diacetyl levels is to increase the rate of utilization of α-acetolactate by accelerating the flux through the pathway leading to the production of the amino acid valine. This has been achieved by transforming yeasts with extra copies of the *ILV5* gene. Such yeasts have

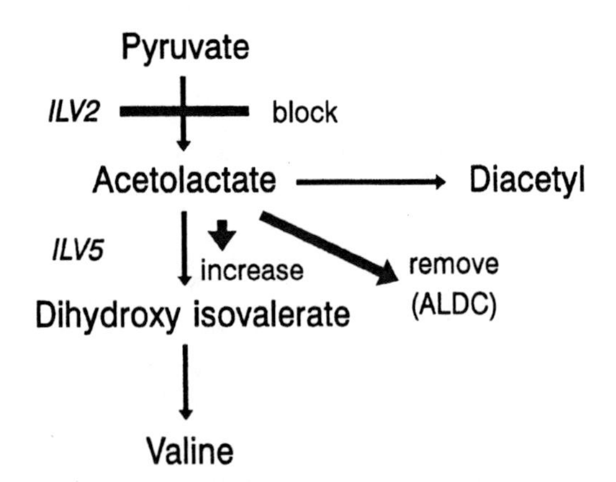

Fig. 7.12 The valine synthetic pathway in *S. cerevisiae* and the formation of diacetyl. The various ways of eliminating diacetyl by reducing the level of α-acetolactate are indicated. Genes: *ILV2*, acetolactate synthase; *ILV5*, reductoisomerase. ALDC; α-acetolactate decarboxylase.

increased α-acetolactate reductoisomerase activity and show a considerable decrease in diacetyl levels (Goossens *et al.*, 1993).

Rapid removal can also be achieved by the use of the enzyme acetolactate decarboxylase (ALDC) which converts α-acetolactate directly to acetoin, by-passing the formation of diacetyl. ALDC is produced by many bacteria and the gene has been cloned from several of them, including species of bacteria used for food production such as *Lactococcus* (Goelling and Stahl, 1988) and *Acetobacter* (Yamano *et al.*, 1994). For approval reasons these are the preferred source of ALDC genes for transformation of brewing organisms. Yeasts transformed with the gene coding for ALDC produce much less diacetyl during fermentation than their untransformed parents (Tada *et al.*, 1995), resulting in a significant saving in maturation time (Figure 7.13).

7.7.3 Yeasts with a reduced capacity to produce hydrogen sulphide

During fermentation many yeasts produce hydrogen sulphide and, like diacetyl, this also has to be removed by a maturation process. A gene has been cloned (*NHS5*, Figure 7.14)) which enhances cystathionine synthesis and so removes excess H_2S which otherwise would escape from yeast into the beer. A brewing yeast transformed with this gene has been described which produces considerably less H_2S than normal untransformed yeasts (Tezuka *et al.*, 1992). An alternative approach to H_2S control has been

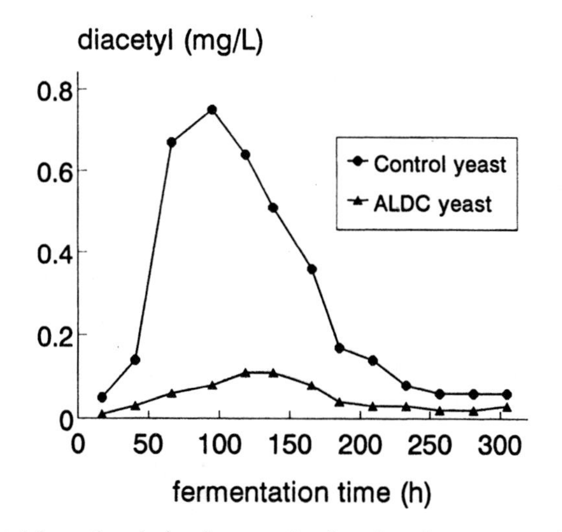

Fig. 7.13 Diacetyl formation during fermentation by a brewing yeast transformed with the ALDC gene from *Acetobacter aceti* ssp. *xylinum*. Fermentations were carried out at 12°C using 40 litres of lager wort. Yeasts were pitched at the rate of 5 g wet weight/litre into pre-aerated wort. Diacetyl levels were monitored by headspace gas-liquid chromatography using an electron capture detector.

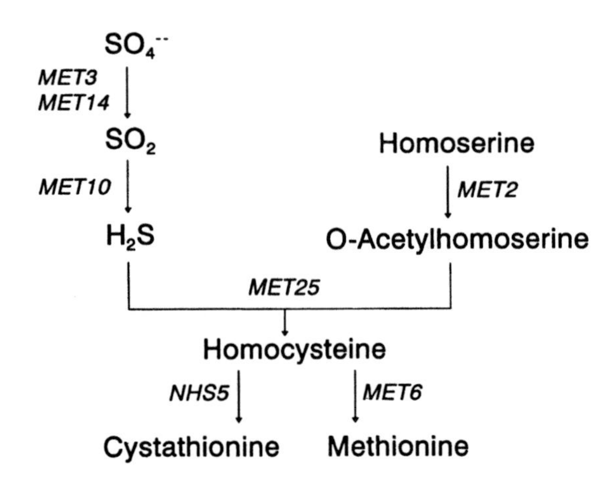

Fig. 7.14 Genes of the sulphate metabolic pathway in *S. cerevisiae*. *MET2*, homoserine acetyltransferase; *MET3*, ATP sulphurylase; *MET6*, homocysteine methyltransferase; *MET10*, sulphite reductase; *MET14*, APS kinase; *MET25*, O-acetylhomoserine, O-acetylserine sulphhydrase; *NHS5*, cystathionine β-synthase.

provided by the cloning of the *MET25* gene (Figure 7.14) (Omura *et al.*, 1995). A brewing yeast transformed with *MET25* showed a level of expression several times higher than that of the parent yeast. In pilot scale fermentations the transformant behaved normally but produced less H_2S than the normal yeast.

7.7.4 *Increased production of flavour compounds*

The level of sulphur dioxide in finished beer is very important for flavour stability since it acts both as a means of sequestering the carbonyl compounds responsible for stale flavours and as an antioxidant. Two different approaches have been used to manipulate sulphur dioxide levels in beer (Figure 7.14, above). Firstly, the *MET3* and *MET14* genes have been cloned and expressed in brewing yeasts. The resulting higher activities of the enzymes coded by these genes increased the flux through the sulphate metabolism pathway and produced higher levels of sulphur dioxide (Korch *et al.*, 1991).

The second strategy involves gene disruption rather than over-expression (Hansen and Kielland-Brandt, 1995). In different yeasts, recombinant DNA techniques have been employed to disrupt the *MET2* and *MET10* genes respectively in order to decrease the synthesis of methionine. The resultant lower concentration of the amino acid would be expected to derepress the sulphate metabolic pathway and hence increase the amount of sulphur dioxide produced. When the *MET10* disruptant was used in fermentations, this was indeed observed and coincidentally the levels of hydrogen sulphide were considerably reduced. With the *MET2* disruptant, although the level of sulphur dioxide did increase, this was accompanied by an increase in hydrogen sulphide levels because of the normal functioning of sulphite reductase (coded by the *MET10* gene) in this strain.

Acetate esters are synthesized in yeast from acetyl-CoA and the relevant alcohols by the action of the enzyme alcohol acetyltransferase (AAT). The gene which encodes AAT (*ATF1*) has been cloned (Fujii *et al.*, 1994) and used in yeast transformation experiments. Transformants carrying multiple copies of the gene exhibited high AAT activity and produced larger amounts of acetate esters than the control yeasts. So far brewing yeasts have not been transformed but now that the gene is available this will undoubtedly soon follow.

7.8 Conclusions

In the laboratory, considerable progress has already been made in producing brewing yeasts modified in a number of very specific ways. The possibilities for enhancing the efficiency of conversion of carbohydrate to ethanol and for producing beers with modified flavours are already the subject of much active research. With a growing understanding of the physiology and genetic make-up of industrial yeasts much more will undoubtedly become possible in the future. Such developments will, in the long term, revolutionize the brewing process, improving efficiency and even leading to new products. There are perceived public acceptability problems in the use of genetically modified yeasts for beer production but the first hurdle has already been passed in obtaining regulatory permission to use such a yeast in at least one country. This is a small but vital step towards general acceptance of the technology.

Acknowledgements

The author thanks the Director General of Brewing Research International for permission to publish.

References

ACGM/HSE/DOE (1993) Guidelines for the risk assessment of operations involving the contained use of genetically modified micro-organisms. *Note 7*, ACGM Secretariat, London.

Andrews, B. J. *et al.* (1985) The FLP recombinase of the 2μ circle DNA of yeast: interaction with its target sequences. *Cell*, 40, 795–803.

Aschengreen, N. H. (1987) Enzyme technology. *Proceedings of the European Brewery Convention Congress, Madrid*, 221–31.

Berghof, K. and Stahl, U. (1991) Improving the filterability of beer by the use of β-glucanase active brewers' yeast. *BioEngineering*, 7, 27–32.

Cantwell, B. A. and McConnell, D. J. (1983) Molecular cloning and expression of a *Bacillus subtilis* β-glucanase gene in *Escherichia coli. Gene*, 23, 211–19.

Department of Health (1991) Guidelines on the assessment of novel foods and processes. *Report on Health and Social Subjects*, 38, HMSO, London.

Enari, T. M. *et al.* (1987) Glucanolytic brewer's yeast. *Proceedings of the European Brewery Convention Congress, Madrid,* 529–36.

Fujii, T. *et al.* (1994) Molecular cloning, sequence analysis, and expression of the yeast alcohol acetyltransferase gene. *Applied and Environmental Microbiology*, 60, 2786–92.

Goelling, D. and Stahl, U. (1988) Cloning and expression of an α-acetolactate decarboxylase gene from *Streptococcus lactis* subsp. *diacetylactis* in *Escherichia coli. Applied and Environmental Microbiology*, 54, 1889–91.

Goossens, E. *et al.* (1993) Decreased diacetyl production in lager brewing yeast by integration of the *ILV5* gene. *Proceedings of the European Brewery Convention Congress, Oslo*, 251–8.

Hammond, J. R. M. (1995) Genetically-modified brewing yeasts for the 21st century. Progress to date. *Yeast*, 11, 1613–27.

Hammond, J. R. M. (1996) Yeast genetics, in *Brewing Microbiology*, 2nd edn. (F. G. Priest and I. Campbell, eds), Chapman & Hall, London.

Hansen, J. and Kielland-Brandt, M.C. (1995) Genetic control of sulphite production in brewer's yeast. *Proceedings of the European Brewery Convention Congress, Brussels*, 319–28.

Henderson, R. C. A., Cox, B. S. and Tubb, R. S. (1985) The transformation of brewing yeasts with a plasmid containing the gene for copper resistance. *Current Genetics*, 9, 133–8.

Jackson, E. A., Ballance, G. M. and Thomsen, K. K. (1986) Construction of a yeast vector directing the synthesis and release of barley 1-3,1-4-β-glucanase. *Carlsberg Research Communications*, 51, 445–58.

Kielland-Brandt, M. C. *et al.* (1989) Yeast breeding. *Proceedings of the European Brewery Convention Congress, Zurich*, 37–47.

Knowles, J. K. C. *et al.* (1985) Transfer of genes coding for fungal glucanases into yeast. *Proceedings of the European Brewery Convention Congress, Helsinki*, 251–8.

Korch, C. *et al.* (1991) A mechanism for sulphite production in beer and how to increase sulphite levels by recombinant genetics. *Proceedings of the European Brewery Convention Congress, Lisbon*, 201–8.

Lancashire, W. E. and Wilde, R. J. (1987) Secretion of foreign proteins by brewing yeasts. *Proceedings of the European Brewery Convention Congress, Madrid*, 513–20.

Livingstone, D. M. (1977) Inheritance of the 2μm DNA plasmid from *Saccharomyces. Genetics*, 86, 73–84.

Meaden, P. G. *et al.* (1985) A *DEX* gene conferring production of extracellular amyloglucosidase on yeast. *Gene*, 34, 325–34.

Mellor, J. *et al.* (1983) Efficient synthesis of enzymatically active calf chymosin in *Saccharomyces cerevisiae. Gene*, 24, 1–14.

Omura, F. *et al.* (1995) Reduction of hydrogen sulphide production in brewing yeast by constitutive expression of *MET25* gene. *Journal of the American Society of Brewing Chemists*, 53, 58–62.

Ono, K. *et al.* (1988) Various molecular species in glucoamylase from *Aspergillus niger. Agricultural and Biological Chemistry*, 52, 1689–98.

Pasteur, L. (1879) *Studies on Fermentation*. Macmillan & Co., London.

Pecquet, S. *et al.* (1991) Kinetics of *Saccharomyces cerevisiae* elimination from the intestines of human volunteers and effect of this yeast on resistance to microbial colonization of gnotobiotic mice. *Applied and Environmental Microbiology*, 57, 3049–51.

Pretorius, I. S. *et al.* (1986) Molecular cloning and characterisation of the *STA2* glucoamylase gene of *Saccharomyces diastaticus. Molecular and General Genetics*, 203, 29–35.

Rose, A. H. and Harrison, J. S. (1987) Introduction, in *The Yeasts*, Vol. 1. *Biology of Yeasts* (A. H. Rose and J. S. Harrison, eds). Academic Press, London.

Tada, S. *et al.* (1995) Pilot scale brewing with industrial yeasts which produce the α-acetolactate decarboxylase of *Acetobacter aceti* ssp. *xylinum. Proceedings of the European Brewery Convention Congress, Brussels*, 369–76.

Tezuka, H. *et al.* (1992) Cloning of a gene suppressing hydrogen sulphide production by *Saccharomyces cerevisiae* and its expression in a brewing yeast. *Journal of the American Society of Brewing Chemists*, 50, 130–3.

Vakeria, D. and Hinchliffe, E. (1989) Amylolytic brewing yeast: their commercial and legislative acceptability. *Proceedings of the European Brewery Convention Congress, Zurich*, 475–82.

Yamano, S., Tanaka, J. and Inoue, T. (1994) Cloning and expression of the gene encoding α-acetolactate decarboxylase from *Acetobacter aceti* ssp. *xylinum* in brewer's yeast. *Journal of Biotechnology*, 32, 165–71.

Yamashita, I., Suzuki, K. and Fukui, S. (1985a) Nucleotide sequence of the extracellular glucoamylase gene *STA1* in the yeast *Saccharomyces diastaticus*. *Journal of Bacteriology*, 161, 567–73.

Yamashita, I. *et al.* (1985b) Polymorphic extracellular glucoamylase genes and their evolutionary origin in the yeast *Saccharomyces diastaticus*. *Journal of Bacteriology*, 161, 574–82.

Yocum, R. R. (1986) Genetic engineering of industrial yeasts. *Proceedings of Bio Expo 86*, 171–80.

8 Baker's yeast

RUTGER VAN ROOIJEN and PAUL KLAASSEN

8.1 Introduction and historical perspective

When the subject of modern biotechnology is brought up, many people think first of pharmaceutical and medical products like interferon, human growth hormones and insulin, produced by genetically modified microorganisms. These products speak to people's imagination: species originally not equipped with the genetic material to produce scarce and expensive therapeutic substances, are now capable of doing so, often in relatively large quantities and at a lower cost. In the past decade, however, the use of recombinant DNA technology has also increased in the agricultural and food processing industries.

A distinction is made here between the addition to food of a product produced by a genetically engineered microorganism, such as chymosin from *Kluyveromyces lactis*, and the addition of a genetically modified organism (GMO) itself, as in the case of baker's yeast. Although a number of genetically modified yeasts have been approved for use in foods in several countries in the early 1990s, none of these strains is currently commercially available in the open market. However, as public attitudes towards the application of GMOs and their products in the food industry develop, it is only a matter of time before such commercial strains become available.

The history of bread may be as old as the known history of mankind. The basic recipe for the preparation of bread has changed very little with time. The discovery of leavened bread in Egypt, thought to have occurred accidentally, may be considered as a revolutionary step in breadmaking (Sheppard and Newton, 1957). This discovery eventually led to the insight that the digestibility and stability, as well as the texture, of the baked product are improved when the dough is left undisturbed for a period of time prior to baking. The use of crude leaven, most probably a mixture of wild yeast(s) and lactic acid bacteria, by different nations and cultures has been reported (Sheppard and Newton, 1957). The best known example is the use of brewer's yeast or barm, collected from the surfaces of fermenting beers and wines, by bakers to make bread. Reliable doughs were uncommon in the past because such barms were weak in fermentation properties.

Industrial baker's yeast production and applications are often considered to have started with the Vienna Process in 1846. This was just before the role of yeasts in fermentation was demonstrated by Pasteur between 1857 and 1863 (Oura *et al.*, 1982). By 1872, several yeast companies were founded and a factory specializing in the production of fresh baker's yeast was established in France. In 1870, J.C. van Marken introduced the Vienna Process in Holland and built a factory, The Dutch Yeast and Spirits Factory, now known as Gist-brocades. Van Marken believed that the success of his company depended on basic research (Bennet and Phaff, 1993). Some of the names associated with this research establishment, amongst others Beijerinck and Kluyver, are well known in microbiology.

At the beginning of the 20th century compressed yeast, specially manufactured for panary fermentation, was introduced. Nowadays, more than half a billion tonnes (dry weight) of baker's yeast is being produced worldwide. Baker's yeast, mostly *Saccharomyces cerevisiae*, as a commercial product is available in three main forms: compressed yeast, dried yeast and cream yeast.

There are three main worldwide players in the field of baker's yeast production: the Dutch company Gist-brocades, the French company 'LeSaffre', and Mauri (Burns Philp, Australia). Baker's yeast companies must be able to adapt to international markets and meet the varied bread-making requirements of many different cultures and nationalities. To this end, the larger yeast companies often form collaborative partnerships with local yeast companies.

8.2 Traditional approaches to yeast improvement

For many years, industrial research has focused on improvement of the price/performance ratio of baker's yeast, that is, developing an optimum quality of yeast at an acceptable cost. A yeast strain has to fulfil a number of criteria determined primarily by market demand but also by production requirements. These criteria are often mutually incompatible: maximum yield, for example, *versus* maximum leavening activity or maximum quality *versus* minimum costs. In practice, commercial yeasts are often a combination of sub-optimal quality characteristics.

The properties of a yeast may be influenced to some extent by varying fermentation parameters such as ammonia feed, growth rate and/or the duration of fermentation. In this way, enzymatic activities as well as the macromolecular composition of the yeast may be modulated. However, the disadvantage of this approach is that other properties of the yeast may be changed as well. In addition, this approach is only possible when certain hereditary properties are already present in the yeast genome. Thus, a yeast lacking a genetic element (gene) encoding the protein involved in the uptake of maltose (the enzyme maltose permease), will under no circumstances be able to utilize this sugar, regardless of the applied fermentation regime.

Several techniques of classical genetics such as mutagenesis by irradiation or chemical agents, have been used in the past to improve baker's yeast. Hybridization (sexual breeding or mating) has been the most widely applied technique. *S. cerevisiae* is a sexual organism that is able to sporulate under certain conditions, giving rise to two spores with different mating types, a and α. Haploid cells with opposite mating types originating from these spores may be recombined to form a new diploid strain. This technique has been extremely useful in the past in the study of yeast genetics and for breeding novel yeast strains. For instance, haploid strains derived from a yeast strain with excellent gassing power but poor drying properties may be crossed with haploid cells from a strain with good drying properties to produce a novel strain with both good gassing power and drying properties.

In practice, the value of hybridization for the construction of new industrial strains has been limited. Firstly, a good screening protocol allowing easy discrimination between improved and unchanged strains for large numbers of newly constructed diploids is often not available. Secondly, many industrial strains sporulate poorly or not at all. Thirdly, and probably most importantly, this technique is not very precise. In a mating experiment, approximately half the genome of one parent strain is mixed with approximately half the genome of the other. Even if a new hybrid with improved characteristics is formed, often other advantageous properties are lost at the same time. This then necessitates several rounds of back-crossing of the new hybrid with one of the parents, in order to retrieve the lost advantageous properties. Clearly, this method of strain improvement is very time-consuming. In the late 1970s, new molecular genetic techniques became available which allowed much more specific manipulations of the yeast genome to be made.

8.3 The construction of genetically modified baker's yeast

Recombinant DNA techniques are now preferred to the hybridization procedure described above for achieving improvements in baker's yeast. The new technology provides the potential for precise genetic modification by introducing one or more extra copies of a gene, deleting one or more genes, changing the regulation of expression of genes, or modifying the structure of a protein (usually an enzyme). Using this technology, it is possible to construct new yeast strains in which the DNA has been 'reshuffled' without introducing any foreign (non-yeast) DNA.

As already noted in Chapter 1, the successful application of recombinant DNA techniques to any organism requires three fundamental steps: transformation (a process for introducing new DNA into cells); a plasmid vector to carry the transforming DNA into the host cell; and a selection method which allows discrimination between transformed and untransformed cells

(Tuite, 1992). Specific methods used for yeast are described in general terms in Chapter 7 and illustrated in Figure 7.1.

8.3.1 *Transformation of yeast*

In the late 1970s, the genetic transformation of yeast protoplasts was first reported (Hinnen, 1978; Beggs, 1978). A major problem with this technique was the relatively high frequency of protoplast fusion, resulting in transformants of higher ploidy than in the original strain thereby hampering the effective evaluation of the transformation process (Harashima *et al.*, 1984). Subsequently, a method which allowed the transformation of whole cells was developed (Ito *et al.*, 1983; Schiestl and Gietz, 1989). It was shown that incubation of yeast cells with alkali metal ions, such as Li^+, and PEG (polyethylene glycol) rendered those cells 'competent', that is able to take up DNA. An advantage of this method, compared with the transformation of protoplasts, was that most strains, including the more difficult industrial strains, could be transformed provided limited modifications to the standard protocols were made.

Alternative transformation methods have been developed including fusion of protoplasts to liposomes carrying transforming DNA (Ahn and Pack, 1985), bombardment of whole cells with microprojectiles coated with transforming DNA (Johnston and Butow, 1988) and electroporation. Transformation of *S. cerevisiae* by electroporation was first achieved with protoplasts (Karube *et al.*, 1985) and subsequently also with intact cells (Hashimoto *et al.*, 1985). Again, the rather low transformation frequencies reported in these early publications were later improved (Becker and Guarente, 1991) and optimized for commercial baker's yeast strains (Gysler *et al.*, 1990).

8.3.2 *The transforming DNA*

In the next step, cloned (duplicated) DNA is introduced into the host cell using plasmid vectors. The plasmids used for *S. cerevisiae* share two basic properties: they are able to replicate in *Escherichia coli* and the selection markers present on these plasmids allow discrimination between transformed and untransformed cells in both *E. coli* and *S. cerevisiae.*

For improving industrial baker's yeast strains, the so-called episomal and integrative/integrating plasmids are the most frequently used vectors. The main features of these vectors are briefly discussed below but for a detailed description of all types of yeast vectors, the interested reader is referred to reviews by Parent *et al.* (1985) and Tuite (1992).

Episomal plasmids are cloning vehicles used for the introduction of extra copies of specific genes into *S. cerevisiae* at a high copy number, for example, 10–30 copies per cell. Episomal plasmids replicate autonomously (i.e.

independently of chromosome replication) in both *E. coli* and *S. cerevisiae* (Murray, 1987). Integrating plasmid vectors are used for stable introduction of genes into the chromosomal DNA of *S. cerevisiae*. Integrating vectors do not replicate in yeast and can only be maintained upon integration into the genome of *S. cerevisiae*.

Deletion of a gene in yeast is achieved firstly by cloning (copying) the target gene into a plasmid vector. Next, another DNA sequence, usually a marker gene, is cloned into the plasmid within the DNA sequence of the target gene. This may be done by cloning the marker in a restriction site (i.e. disruption of the coding sequence), or by replacing all or a part of the coding sequence by the marker (gene-replacement). The whole plasmid or a part of it is then offered to competent cells and integrated by homologous recombination (Rothstein, 1983).

To increase the expression level of a gene in industrial baker's yeast, episomal plasmids carrying the target gene have been used in the past. In this way, a high copy number, and thus a high gene dosage was achieved. More recently, integrative plasmids containing the target gene together with a strong constitutive promoter have been developed. A promoter is a DNA sequence which controls the expression of the coding sequence it precedes. By using a strong and constitutive promoter, a gene is expressed at a high level, practically independently of culture conditions. The advantages of integrative over episomal plasmids are that integration into the chromosome leads to more stable transformants and, in general, up to five copies of the plasmid may be integrated into the host cell's genome. This allows the evaluation of the effect of the gene copy number on the performance of the recombinant DNA yeast. Nowadays, many standard vectors containing strong constitutive promoters are available, into which any gene may be cloned.

Since the late 1980s, the polymerase chain reaction (PCR) (Saiki *et al.*, 1985, 1988) has become a very powerful tool for molecular biologists so that much gene cloning may be reduced or even omitted. It is now possible to disrupt genes in one step, after a simple PCR reaction in a test tube (Wach, 1996). Specific examples of the first two methods of genetic modification described above are discussed in section 8.4.

8.3.3 *Selection of transformants*

Antibiotic resistance markers (e.g. conferring resistance to gentamycin, phleomycin or hygromycin B) are often used in industrial yeast strains to allow rapid and efficient selection of transformants. Although other markers such as the ability to utilize a particular nutrient might be desirable, these are often not available for industrial strains due to their diploid, or sometimes even polyploid or aneuploid, nature. Nevertheless, alternative marker systems are currently being developed. For example, a marker system from *Aspergillus nidulans* based on the ability to use acetamide as nitrogen source has been developed (Selten *et al.*, 1993).

8.3.4 Self-cloning in yeast

An important issue in the genetic modification of industrial baker's yeast is that the final product should be devoid of any gene or DNA sequence not belonging to *Saccharomyces* species. Therefore, techniques have been developed for constructing the so-called 'homologous' recombinant DNA strains (self-cloning). For example, a patent (Osinga, 1987) describes yeast strains with an enhanced ability to utilize the disaccharide maltose, even under conditions where the wild-type yeast is unable to do so, by inserting extra copies of the so-called *MAL* genes. The extra copies of these genes (*MAL6T* and *MAL6S*), have been introduced into a silent, non-essential gene, called *SIT4* (sporulation induced transcript, Gottlin-Ninfa and Kaback, 1986). The use of this silent gene together with a series of transformation steps and serial growth on media containing acetamide as sole nitrogen source allows for the positive selection of only those transformants containing homologous genes (Needleman, 1991).

8.4 Technological properties of genetically modified baker's yeast strains

Three examples of genetically modified baker's yeast strains with improved properties are presented in this section. The technological advantages and disadvantages of these strains are discussed by comparison with their wild-type relatives.

8.4.1 Increased expression of MAL *genes*

During leavening, baker's yeast converts the sugars present in the dough into carbon dioxide. Sugared doughs contain the easily digestible sucrose while maltose, derived from flour starch, is the most abundant sugar in lean dough. Usually, each type of dough requires its own type of baker's yeast strain. On the one hand, strains of yeast with the ability to adapt rapidly to maltose often perform poorly in sugared doughs; on the other hand, sugar-resistant yeast strains generally perform relatively poorly in lean doughs. This is because flour contains low levels of hexoses, such as glucose, which inhibit the expression of the two genes responsible for maltose metabolism in sugar-resistant yeast: the *MAL6T* gene encoding maltose permease which transports maltose across the plasma membrane into the cell, and the *MAL6S* gene encoding maltase which hydrolyses maltose intracellularly. Therefore, extra copies of the *MAL* genes were introduced into sugar-resistant yeast strains with the principal aim of making these strains suitable for

Table 8.1 Relative gas production in lean dough and keeping quality of industrial baker's yeast Strain A and its genetically modified derivatives with one or two altered *MAL* genes. Gas values have been corrected to 285 mg dry matter

Industrial strain code	Constitutive *MAL* genes	Relative gas production (%)	Keeping quality (% leavening activity remaining after 4 d at 23°C)
A	none	100	90
A pGB-iA32/G418	*ADH1/MAL6T* (maltose permease)	111	91
A pGb-iRR01	*ADH/MAL6T* and *TEF/MAL6S* (maltose permease and maltase)	133	88

use in both lean and sugared doughs. In addition, the natural promoters of these genes were replaced by both strong and constitutive promoters, in order to abolish the natural regulation (and therefore repression by glucose) of these genes (Osinga, 1987).

Industrial strain A was transformed with integrating vectors, containing one or two *MAL* genes. Parental strain A and two integrative transformants were grown in a fermenter at lab scale in a downscaled process derived from the industrial production process. The gas production ability of the yeast was determined immediately after harvest and after storage at 23°C for 4 days. Table 8.1 shows that integration of extra copies of the maltose permease gene *ADH1/MAL6T* into the chromosome improved gas production in lean dough significantly. When both an altered maltose permease gene and an altered maltase gene were integrated in the chromosome of commercial strain A, gassing power was improved even further. In a typical experiment, about 30% more CO_2 was produced after 165 minutes in a lean dough, corresponding to a level of about 410 ml CO_2 per 285 mg dry weight of yeast. Furthermore, the improvement in leavening activity was maintained during storage at 23°C. The extent of loss of leavening activity during storage was practically identical for the parental and the novel strains (Table 8.1).

The novel strains described above contained heterologous DNA, that is DNA originating from species other than *S. cerevisiae*. Consequently, further efforts were made to construct strains containing only DNA derived from *S. cerevisiae*. Parental strain A was genetically modified in such a way that *ADH1/MAL6T* and *TEF/MAL6S* were introduced via the silent *SIT4* gene by the technique of gene replacement described in the previous section. The parental strain A and the genetically modified derivative designated A pGb-p2RBRR01#1 were grown on molasses similarly to the commercial aerobic fermentation. Gas production in a standard lean dough test was 18% higher when using the genetically modified organism, compared with the control. Although substantial, this improvement was not as great as the 30% improvement obtained using the genetically modified yeast strain containing heterologous DNA (Table 8.1, above). This difference in performance was attributed to the greater number of copies of the integrative plasmid in the genome of the heterologous strain compared with the homologous strain. Subsequently, by further increasing the copy number of altered *MAL* genes, a new homologous yeast strain was obtained with similar gas production levels as the heterologous DNA containing yeast strain.

As shown above, by using recombinant DNA techniques, a multi-purpose yeast strain has been constructed, designated as *MAL* yeast, which performs almost equally well in lean doughs as in 20% sugar doughs. A multi-purpose yeast has advantages for the baker producing several types of fermented dough systems. Instead of bothering about which yeast has to be used in which dough, one yeast is suitable for all doughs. In addition, instead of producing two different types of yeast, one for the lean dough market and one for the sugared dough market, the yeast producer need manufacture only one yeast strain, leading to a reduced cost-price for the product. The novel *MAL* yeast strain also had the added advantage of producing more gas in less time in lean and sugared doughs thereby offering the baker the prospect of reduced leavening times.

No clear disadvantages (like the taste and texture of the bread) have been observed as a result of the introduction of the constitutively expressed *MAL* genes. It was feared that a negative effect on the dryability of the product would be observed. The constitutive expression of maltose permease might have led to higher amounts of this protein in the plasma membrane, resulting in a lower stability of the membrane. Consequently, a relatively lower gassing power of the dried yeast product might have been observed. However, this appeared not to be the case. The dryability of the *MAL* yeast was comparable to that of the untransformed parental strain.

8.4.2 Yeast with an enhanced trehalose content

From production to its application in dough, yeast is subjected to a wide range of environmental stresses, for example, osmotic stress (high sugar concentrations in dough, salting during downstream processing), nutrient starvation (during storage of fresh yeast), drying and/or freezing (frozen dough application). It has been reported by many researchers (reviewed by Wiemken, 1990) that trehalose may serve as a general stress protectant. Therefore, a baker's yeast strain with a higher trehalose content would be expected to perform better in a dough system in which the yeast experiences some kind of stress. In order to achieve this, the gene most likely to be responsible for the degradation of trehalose (i.e. encoding the enzyme trehalase) was cloned (Driessen *et al.*, 1990; Klaassen *et al.*, 1994). With this cloned DNA, an integrating plasmid was constructed with which the

Table 8.2 Neutral trehalase activity in industrial baker's yeast Strain A and its genetically modified derivative (ApTRE-dis)

Culture conditions	Strain	Neutral trehalase activity (mU/mg protein in cell free extract)
Batch culture 2.5% glucose	A	26.1
	A pTRE-dis	12.2
Batch culture 2.5% glycerol and 0.2% glucose	A	27.7
	A pTRE-dis	14.7
Continuous culture (Growth rate = 0.06 h^{-1})	A	11.6
	A pTRE-dis	4.8
Continuous culture (Growth rate = 0.15 h^{-1})	A	7.9
	A pTRE-dis	3.3

chromosomal copy of the gene was inactivated. Integration of this construct into the chromosome at the neutral trehalase locus led to two nonfunctional copies of the gene, one lacking the 5′ end of the gene, the other the 3′ end of the gene. Two strains were transformed in this way: industrial strain A, and the diploid strain G.

Analysis of the transformants of strain A revealed that neutral trehalase activity was reduced by approximately 50% under four different conditions of cultivation including batch-wise and continuous fermentation, as shown in Table 8.2. The effect of disruption of one copy of the gene encoding neutral trehalase on the trehalose content of the cells in the stationary phase was also studied. Strains A, G and their respective transformants were grown overnight in batch culture and the neutral trehalase activity and trehalose content were determined. As shown in Table 8.3, a lower neutral trehalase activity resulted in an increased trehalose content.

Increased trehalose content has, in turn, been shown to have a positive effect on the shelf-life of baker's yeast. A higher resistance to drying and osmotic stress was also found to be related to a higher trehalose content. During drying of yeast, the normal bilayer structure of the membranes is destabilized by the extraction of water. When the cells are rehydrated, the membrane is vulnerable and essential components leak from the cell, which

Table 8.3 Neutral trehalase activity and trehalose content of baker's yeast strain A, G and their genetically modified derivatives with one inactive neutral trehalase gene

Strain	Neutral trehalase activity	Trehalose content (%, w/w)
Industrial strain A	4.4	5.7
A pTRE-dis	3.0	7.9
Diploid strain G	1.3	2.6
G pTRE-dis	0.2	3.5

may influence dough leavening negatively. Leakage may be reduced by stabilization of the membrane. This may be achieved by increasing the intracellular trehalose content. It has been shown that the binding of trehalose to the phospholipid fraction lowers the phase transition temperature of the membrane (Leslie *et al.*, 1994). In addition, during high external osmotic pressures (as in sugar doughs) a high(er) trehalose level is beneficial for the yeast as it raises the intracellular osmotic value. Consequently, less water is lost from the yeast resulting in a relatively better leavening activity under high osmotic pressures. Finally, yeast having an increased trehalose content appeared to be more resistant to freeze stress as it occurs in frozen dough applications.

8.4.3 Introduction of futile cycles in the glycolytic pathway

One method of improving the leavening activity of baker's yeast is the introduction of futile cycles in the metabolism of *S. cerevisiae* (Rogers and Szoztak, 1987; Navas *et al.*, 1993; van Rooijen *et al.*, 1994). A futile cycle is the simultaneous occurrence of a catabolic and a corresponding anabolic biochemical reaction, in which ADP and ATP are generated, resulting in the net hydrolysis of ATP. In this way, the cellular ATP level is reduced, thereby stimulating glycolysis, presumably by stimulating those glycolytic steps in which ATP is generated. A higher glycolytic flux results in a higher level of CO_2 production.

In a Gist-brocades patent application, two genes encoding gluconeogenic enzymes were constitutively expressed: the *FBP1* gene encoding fructose-1,6-bisphosphatase, and the *PCK1* gene, encoding phosphoenolpyruvate carboxykinase. Normally, the genes encoding these gluconeogenic enzymes are not transcribed during growth on glucose. The promoter regions controlling the transcription of the gluconeogenic genes were replaced by both strong and constitutive promoters, thereby allowing expression of the gluconeogenic enzymes under all conditions. The concept was initially tested in laboratory strains but has since been proven successful in commercial baker's yeast strains as well.

The effects of two futile cycles on yeast resistance to sugar, calcium propionate and sodium chloride were tested. The results, presented in Table 8.4, clearly show that the positive effect of futile cycles increased with increasing 'stress'. In a dough containing 3% sucrose, no advantageous effects were observed from the introduction of two futile cycles. However, in a dough containing 20% sucrose, the strain containing two futile cycles produced 19% more gas compared with the parental strain. Increasing the stress conditions even further by the addition of calcium propionate and sodium chloride demonstrated even greater advantages of the futile cycle stimulation factor in yeast (Table 8.4).

Table 8.4 Relative CO_2 production after 165 minutes by baker's yeast Strain CJM152 and its genetically modified derivative CJM186 in 3% and 20% sugar doughs with and without added calcium propionate (CaP, 0.3%) or sodium chloride (NaCl, 2%)

Sucrose added to the dough (%)	Additional ingredients	Relative gas production by control strain CJM152 (%)	Relative gas production by modified strain CJM186 (%)	Futile cycle stimulation factor (%)*
3	–	100	100	100
3	CaP	84	99	117
3	NaCl	20	73	365
20	–	36	43	119
20	CaP	50	106	212
20	NaCl	4	15	375

* The futile cycle stimulation factor (FCSF) was calculated by dividing the relative amount of CO_2 produced in the control strain CJM152 by the relative gas production of the strain expressing two futile cycles, CJM186. A FCSF exceeding 100 indicates that the presence of two futile cycles stimulates gas production by the percentage given.

Since dried yeast is an important product, an important prerequisite for the application of 'futile cycle' technology is that the positive effects described above in fresh yeast are retained after drying. Therefore, compressed yeast samples of strains CJM189 (another wild-type strain) and CJM186 (containing the two futile cycles) were dried, and the gas production was determined under various stress conditions. The positive effects of the presence of two futile cycles are clearly shown in Table 8.5. As in compressed yeast, the impact of futile cycles on gas production in dried yeast increased with increasingly stressful conditions. Surprisingly, the dryability of strain CJM186 also increased by 30%.

In general, the gassing power of baker's yeast containing a futile cycle strongly increased with increasing conditions of physiological stress, such as

Table 8.5 Gas production by two dried genetically modified baker's yeast Strains CJM186 and CJM189 in various doughs. The figures are given as percentages of gas production by a compressed yeast sample of strain CJM189 in dough containing 20% sucrose. Under these conditions, this strain produced 207 ml CO_2 per 600 mg dry matter in 165 minutes

Conditions	Strain CJM189	Strain CJM186	Futile cycle stimulation factor (%)*
Compressed, 20% sucrose	100	121	121
Dried, 20% sucrose	37	58	157
Dried, 20% sucrose, 2% NaCl	24	49	204
Dried, 20% sucrose, pH 3.5, 0.3% calcium propionate added	19	44	231
Dryability	37	48	130

* The futile cycle stimulation factor was calculated as indicated in Table 8.4

the addition of relatively high concentrations of sodium chloride, sucrose and calcium propionate, but also with low pH, nutrient starvation, high temperature, high ethanol concentration or low water activity. The main advantage of the genetically modified yeast is the ability to perform under conditions normally considered to be severe. Under these conditions, the genetically engineered yeast produces amounts of CO_2 that are higher than those achievable by the wild-type yeast strain. Consequently, the time required to leaven a dough to a certain height was shorter in the presence of the recombinant yeast. Further studies are currently underway to unravel the exact role of futile cycles in the observed increase in stress resistance.

8.5 Future prospects

In the past, baker's yeast strains with improved characteristics were obtained mainly by classical genetic breeding and selection. Together with optimization of cultivation and downstream processing methods, this has led to a gradual improvement of specific important characteristics like leavening power and shelf-life. In principle, modern genetic techniques are being used to achieve the same goals, but with several advantages. Firstly, the expression of existing genes can be modulated with greater precision and prediction. This modulation can take the form of inactivation or overexpression of a gene. By contrast with classical genetic breeding, the rest of genetic material of the organism is not affected when using the novel recombinant methods. Secondly, copies of genetic material from unrelated species can be introduced, hitherto impossible to achieve by traditional breeding or selection techniques.

One of the main disadvantages of the random approach in classical breeding is that it has resulted in a relatively small population of commercially used baker's yeast strains. Furthermore, the molecular basis of the various improvements achieved by classical breeding is relatively poorly understood. As a consequence, concepts to further significantly improve baker's yeast quality are relatively scarce. This relative lack of understanding is attributed to the points described below.

Manufacturers of baker's yeast generally regard the funding of fundamental research programmes as having a negative costs/profits balance. With a few exceptions, this has resulted in a strategy adopted by many manufacturers of 'just' carefully following the literature and research of the universities and research institutes and testing the hypotheses generated using laboratory strains against the genetic background of industrial baker's yeast. So far, this approach has been relatively successful and has resulted in new strains with potentially additional commercial value.

However, most academic research on baker's yeast is carried out with genetically easily accessible laboratory strains. The prototrophic, aneuploid, hard-to-sporulate industrial strains are not very attractive organisms for most researchers. Because of the differences in the genetic make-up between lab strains and industrial strains it remains to be determined whether extrapolation of physiological or genetic phenomena from lab strains to industrial strains is valid. There are many physiological differences between laboratory and industrial yeast strains including, for example, the ability of industrial strains to produce higher trehalose levels (up to 20% on dry matter), a faster glycolytic flux in industrial strains under both aerobic and anaerobic conditions, the ability of industrial strains to produce other metabolites at the onset of the 'short-term' Crabtree-effect (Niederberger *et al.*, 1995), and finally, the higher maximal growth rate of industrial strains. The presently used commercial strains have evolved in an industrial environment for almost a century, with a maximal glycolytic flux (both aerobic and anaerobic) as one of the main selective pressures for 'survival'. Therefore, the study of the molecular mechanisms underlying the differences between lab and industrial strains could lead to useful new insights.

The physiological conditions applied in a commercial baker's yeast fermentation generally differ from those used in the yeast genetics labs. Baker's yeast is produced by fed-batch fermentation, a heavily aerated process with ammonia/urea and molasses as the main nutrients. Subsequently, the yeast has to perform anaerobically in a dough environment. By contrast, the cultivation conditions applied for the characterization of genes in the specialized genetics labs usually consist of a semi-aerated batch culture with various carbon sources followed by an additional induction phase in other media. Since yeast composition depends on cultivation conditions, the response of the cell to various external stimuli or conditions may vary considerably. In the last decade, efforts have been made to combine the strengths of the physiologically-orientated labs with those of the genetically-orientated labs and to characterize the function of genes under various fixed growth rates in the chemostat (continuous culture). The need for such collaborative efforts has been recognized by the European Commission and has resulted in the start of the 'Cell factory' (Framework 4) programme of research involving many excellent European research groups.

The challenge for the future will be to further establish and intensify the integration between molecular genetics and physiology labs. Only with the knowledge generated from this integrative research and that generated within the EUROFAN project (see below), can new concepts for further improvement of industrial baker's yeast strains be expected.

8.5.1 The impact of genome sequencing projects

In 1996, an historical event took place: the determination of the entire nucleotide sequence of the *S. cerevisiae* genome was completed (Goffeau *et al.*, 1996). With the completion of this project, a new programme of research called EUROFAN was started, and this is now fully operative. The aim of this new project is to discover the function of 1000 genes whose functions are currently unknown. This project is expected to result in an increased understanding of yeast metabolism in general, and regulatory phenomena in particular. It is expected that this increased understanding will, in turn, lead to new ideas for the improvement of a range of processes including the production of baker's and brewer's yeast, heterologous proteins and yeast extracts.

8.5.2 Metabolic pathway engineering

A relatively large proportion of research in the biotechnological industries can be classified as 'metabolic pathway engineering' or the deliberate redirection or velocity modulation of metabolic pathways. Recent examples of the successful redirection of pathways other than those already discussed in section 8.4 are the removal of diacetyl to shorten the lagering time of beer (Farfan *et al.*, 1995) and the synthesis of xylitol from xylose by introduction of the xylitol dehydrogenase gene from *Pichia* (Penttilä and Hann-Hägerdal, 1995). However, the engineering of the *rate* of fluxes of *existing* pathways has been relatively less successful.

In the early days of modern molecular genetics, many projects aiming at improving the flux rate of existing pathways were focused on modulating selected 'rate-limiting' enzymes. However, it was soon discovered that overexpression of genes encoding a selection of rate-limiting enzymes failed to modify the flux rates of pathways (Schaaf *et al.*, 1989). With the introduction of metabolic control analysis (MCA) by the late H. Kacser, it became apparent that this simple view needed to be adjusted or refined. The control over a pathway is now known to be distributed over the entire pathway (Kacser and Burns, 1973; Westerhof, 1995). This will make it difficult to increase the flux of an entire pathway by 'just' modifying the expression of certain key enzymes. The only possibility in the future to improve the 'steady state' flux of pathways will be to first fully understand the physiological driving forces behind the given flux of a certain pathway. An example of the beginnings of such an approach has been described previously in this chapter in which the driving force behind the anaerobic fermentation (i.e. producing ATP for growth) of baker's yeast in doughs was modified by the introduction of ATP-spoiling reactions (section 8.4.3). This resulted in an increase in gas production in doughs in which the driving force for the anaerobic fermentation has been decreased as a consequence of the addition of certain components (such as extra sugar, salt, etc.). However, the maximal fermentation rate in doughs in which this driving force was not or was only slightly affected, could not be increased.

The challenge for future research on baker's yeast will be to identify and fully understand the regulatory phenomena involved in the control of fluxes in metabolic pathways. This fundamental knowledge will be an essential prerequisite for the introduction of new baker's yeast products with improved quality onto the market.

References

Ahn, J. S. and Pack, M. Y. (1985) Use of liposomes in transforming yeast cells. *Biotechnol. Lett.* 7, 553–6.

Becker, D. M. and Guarente, L. (1991) High-efficiency transformation of yeast by electroporation. *Meth. in Enzymol.* 194, 182–87.

Beggs, J. D. (1978) Transformation of yeast by a replicating hybrid plasmid. *Nature* 275, 104–9.

Bennett, J. W. and Phaff, H. J. (1993) *Early Biotechnology: The Delft Connection. ASM News*, Vol. 59, No. 8, 401–4.

Botstein, D. and Fink, G. R. (1988) Yeast: an experimental organism for modern biology. *Science*, 240, 1439–43.

Burke, T. D., Carle, G. F. and Olson, M. V. (1987) Cloning of large-segments of exogenous DNA into yeast by means of artificial chromosome vectors. *Science*, 236, 806–12.

Burke, D. C. (1995) Genetic manipulation: public opinion, political attitudes and commercial prospects – an introductory lecture. *Journal of Applied Bacteriology Symposium (*Supplement) 79, 1S–4S.

Driessen, M., Osinga, K.A . and Herweijer, M. A. (1990) Transformed yeast having increased trehalose content. EP 451896, AU 9173782, NO 9101232, CA 20393.

Farfán, M. J., Aparicio, L. and Calderón, I. L. (1995) Induced expression of a mutant allele that deregulates threonine biosynthesis in *Saccharomyces cerevisiae. European research conference. Control of metabolic flux: metabolic pathway engineering in yeasts.* 7–12 April; Granada, Spain.

Goffeau, A., Oliver, S., Dujon, B. *et al.* (1996) *Final conference of the yeast genome sequencing network* Trieste, Italy.

Gootlin-Ninfa, E. and Kaback, D. B. (1986) Isolation and functional analysis of sporulation induced transcribed sequences from *S. cerevisiae. Mol. Cell. Biol.* 6(6), 2185–97.

Gysler, C., Kneus, P. and Niederberger, P. (1990) Transformation of commercial baker's yeast-strains by electroporation. *Biotechnol. Techn.*, Vol. 4 No. 4, 285–90.

Hammond J. R. M. (1994) In F. G. Priest and I. Campell (eds), *Proceedings of the 4th Aviemore Conference on Malting, Brewing and Distilling.* Institute of Brewing, London, pp. 85–99.

Hammond J. R. M. (1995) Genetically-modified brewing yeasts for the 21st century. Progress to Date. *Yeast* 11, 1613–27.

Hammond, J. and Bamforth, C. (1994) Practical use of gene technology in food production. *The Brewer* Feb. 65–69.

Harashima, S., Takagi, A. and Oshima, Y. (1984) Transformation of protoplasted cells is directly associated with cell fusion. *Mol. Cell. Biol.* 4, 771–8.

Heijs, W. J. M., Midden, C. J. H., and Drabbe, R. A. J. (1994) Biotechnology: attitudes and influencing factors, Eindhoven University of Technology; ISBN 90-6814-047-7.

Hinnen, A., Hicks, J. B. and Fink, G. R. (1978) Transformation of yeast. *Proc. Natl. Acad. Sci. USA* 75:1929–33.

Hurley, R., R., de Louvois, J. and Mulhall, A. Yeasts as human and animal pathogens. *The Yeasts* Vol. 1, 207–81.

Ito, H. *et al.* (1983) Transformation of intact yeast cells treated with alkali cations. *J. Bacteriol.* 153, 163–8.

Johnston, S.A. and Butow, R.L. (1988) Mitochondrial transformation in yeast by bombardment with microprojectiles. *Science*, 240, 1538–41.

Kacser, H. and Burns, J. A. (1973) in *Rate Control of Biological Processes* (ed. D. D. Davies), Cambridge University Press, pp. 65–104.

Klaassen, P. *et al.* (1994) Effects of the NTH1 gene disruption in baker's yeast. *Folia Microbiologica* 39(6), 524–6.
Leslie, S. B. *et al.* (1994) Trehalose lowers membrane phase transitions in dry yeast cells. *Biochim. Biophys. Acta* 1192, 7–13.
Mikuss, M. D. and Petes, T. D. (1985) Recombination between genes located on non homologous chromosomes in *S. cerevisiae Genetics* 101, 369–404.
Murray, J. A. H (1987) Bending the rules: the 2μ plasmid of yeast. *Mol. Microbiol.* 1, 1–4.
Needleman, R. (1991) Control of maltase synthesis in yeast. *Mol. Microbiol.* 5(9), 2079–84.
Navas, M. A., Cerdan, S. and Gancedo, J. M. (1993) Futile cycles in *S. cerevisiae* strains expressing gluconeogenic enzymes during growth on glucose. *Proc. Natl. Acad. Sci. USA* 90, 1290–94.
Niederberger, P. *et al.* (1995) Flux studies at the pyruvate bypass for development of industrial strains of *S. cerevisiae*. Granada, Spain; Control of Metabolic Flux: Metabolic Pathway Engineering in Yeasts.
Osinga, K. A. *et al.* (1987) New recombinant yeast strains. EP 306107, NO 8803919, AU 8821868, DK 88049.
Oura, E., Suomalainen, H. and Viskari, R. (1982) Breadmaking. in *Economic Microbiology* (A. H. Rose, ed.) Vol. 7, 87–146.
Parent, S. A., Fenimore, C. M. and Bostian, K. A. (1985) Vector systems for the expression, analysis and cloning of DNA sequences in *S. cerevisiae. Yeast* 1, 83–138.
Penttilä, M. and Hahn-Hägerdal, B. (1995) Xylose utilization by recombinant *Saccharomyces cerevisiae. European research conference. Control of metabolic flux: metabolic pathway engineering in yeasts.* 7–12 April; Granada, Spain.
Rogers, D. T. and Szoztak, J.W. (1987) Yeast strains. International Publication number: WO 87/03006.
Rooijen, R. J. van, Schoppink, P. J. and Bannkreis, R. (1994) Improvement of gas and alcohol production by yeast. Patent Applications EP 0645094, AU 9474174.
Rothstein, R. J. (1983) One-step gene disruption in yeast. *Meth. in Enzymol.* Vol. 101, 202–11.
Saiki, R. K. *et al.* (1985) Enzymatic amplification of β-globin genomic sequences and restriction site analysis for diagnosis of sickle cell anemia. *Science* 230, 1350–4.
Saiki, R. K. *et al.* (1988) Primer-directed enzymatic amplification of DNA with a thermostable DNA polymerase. *Science* 239, 487–91.
Schaaf, I., Heinisch, J., and Zimmermann, F. K. (1989) *Yeast* 5, 285–90.
Schiestl, R. H. and Gietz, R. D. (1989) High efficiency transformation of intact yeast cells using single stranded nucleic acids as a carrier. *Curr. Genet.*, 16, 339–46.
Selten, G. C. M., Swinkels, B. W. and Gorcom, R. F. M. van (1993) Selection marker gene free recombinant strains, esp. filamentous fungi, and methods for obtaining them. EP 635574, AU 9468628.
Sheppard, R. and Newton, E. (1957) *The Story of Bread.* Routledge & Kegan Paul, London, UK.
Tuite, M. F. (1992) Strategies for the genetic manipulation of *S. cerevisiae. Critical Reviews in Biotechnology*, 12, 157–88.
Wach, A. (1996) PCR-synthesis of marker cassettes with long flanking homology regions for gene disruptions in *S. cerevisiae. Yeast*, 12, 259–65.
Watson, J. D. *et al.* (1987) In: Molecular Biology of the Gene, Vol. 1, The Benjamin/Cummings Publishing Company, Inc.
Westerhof, H. V. (1995) Subtlety in control – metabolic pathway engineering. *Trends in Biotech.* 13, 242–5.
Wiemken, A. (1990) Trehalose in yeast, stress protectant rather than reserve carbohydrate. *Antonie van Leeuwenhoek*, 58, 209–17.

9 Starter cultures for the dairy industry

COLIN HILL and R. PAUL ROSS

9.1 Introduction

Starter cultures are an essential component of all fermented dairy foods including cheese, yoghurt, sour cream and lactic butter. The primary function of these bacteria is the conversion of lactose and other sugars in milk to lactic acid. This acidification contributes to a preservative effect with the result that many pathogenic and spoilage bacteria are inhibited. The associated drop in pH also results in the loss of water from the curd as whey. In addition, starters are responsible for the production of a variety of secondary metabolites, including a number of compounds which are necessary for flavour development.

Starter cultures used in dairy fermentations typically belong to a family of bacteria collectively known as the lactic acid bacteria (LAB). This term describes a collection of functionally-related organisms composed of up to 12 different genera. They are all fermentative, non-motile gram-positive bacteria and generally lack a functional catalase. Those of interest in dairy fermentations belong to the genera *Lactococcus*, *Leuconostoc*, *Streptococcus*, *Enterococcus* and *Lactobacillus*. Traditionally, these organisms were classified based on traits such as carbohydrate metabolism, growth temperature and serological typing. However, recent developments in taxonomic methodologies, such as the use of electrophoretic whole-cell protein fingerprints, have offered closer insights into the evolutionary relationship between different members of the group (Pot *et al.*, 1994).

With regard to their application as dairy starters, the LAB can be subdivided into mesophilic and thermophilic cultures with temperature optima of approximately 26°C and 42°C, respectively. The main mesophilic cultures belong to *Lactococcus lactis* subspp. *lactis* and *cremoris* which are used in the production of such products as Cheddar and Gouda cheese, cultured buttermilk and sour cream. In addition, Cit+ lactococci (biovar. *diacetylactis*) and *Leuconostoc* species are used for the enhancement of flavour in some dairy products. The main thermophilic cultures employed include *Streptococcus thermophilus, Lactobacillus helveticus, Lactobacillus delbrueckii* subsp. *lactis*, and *Lactobacillus delbrueckii* subsp. *bulgaricus* which are used in the production of yoghurt and high scald cheeses including Emmental, Gruyere and Italian cheeses.

Historically, starter cultures were used by man to produce soured cream and cheese long before any bacterial involvement was suspected. In these artisanal operations, product from a previously successful fermentation was used to 'start' subsequent processes. So-called 'mixed strain starters' composed of undefined mixtures of lactic acid bacteria and/or other bacteria have their origins in this approach. The performance of these starters, however, can be variable as the microbial population of the starter continues to evolve. For example, dominance of particular stains in the mix over time can be problematic if the dominant strains produce off-flavours or become susceptible to attack by bacteriophage (bacterial viruses, Figure 9.1). Since the exact composition of 'mixed strain starters' is not known, such problems can be difficult to correct. For this reason, most fermented dairy foods are now manufactured using a mixture of between two and six defined strains. In many cases, these particular strains were selected from mixed strain starters on the basis of their ability to produce acid, their

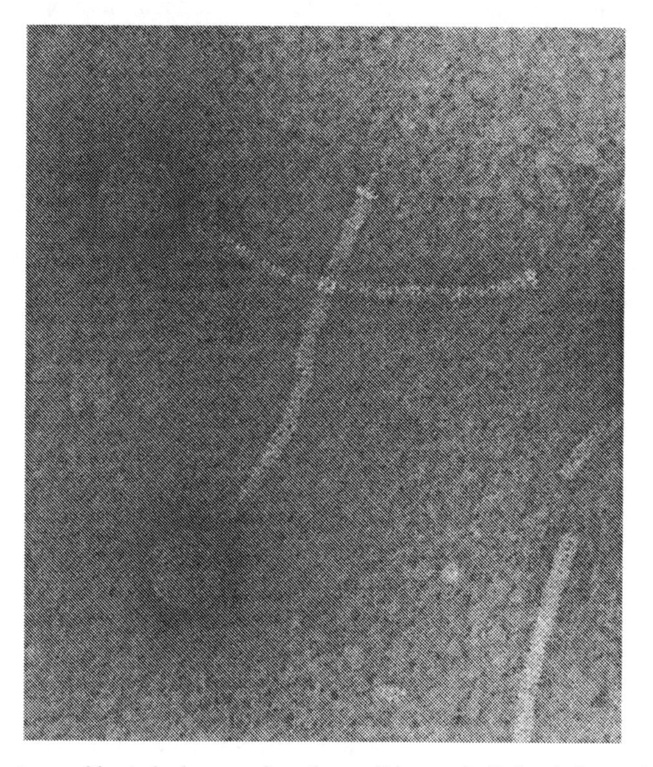

Fig. 9.1 Lactococcal bacteriophage, such as the small isometrically-headed type shown here, are one of the major causes of dairy starter failure in food fermentations. Through genetic approaches valuable industrial starters can now be given protection against such bacteriophage using a range of defence systems which target different stages of the life cycle of the virus.

insensitivity to bacteriophage (phage) or some other desirable trait for commercial manufacture of dairy products. The mixture is normally composed of a set of phage-unrelated strains such that sufficient acid development takes place even in the event of one of the strains succumbing to phage attack. In the event of one starter strain being 'phaged out', it can be replaced by either a phage-unrelated strain or a phage insensitive variant of itself. Alternatively, phage-unrelated defined strain starters can be used on a rotational basis which limits the build-up of specific phage types over time. The result of using defined strains in such approaches has aided the large-scale manufacture of products such as Cheddar and Dutch-type cheese and has meant that rigorous manufacturing time schedules can be met while ensuring that product quality remains optimized.

As a consequence of the economic importance of the LAB, an intensive worldwide research effort has been directed at a thorough understanding of the genetics and metabolism of starter bacteria, particularly the genus *Lactococcus* (reviewed in Venema *et al.*, 1996). Such research has led to the development of efficient approaches for genetic manipulation of industrially significant traits, in addition to providing insights into gene structure and regulation in these organisms. The result of this intense effort has meant that many of the LAB, ultimately destined for the food chain, can now be improved genetically with regard to a number of commercially significant traits. This chapter will address the progress made in the genetic analysis of starter bacteria and discuss how the knowledge base generated is currently being applied to their genetic improvement for specialized industrial uses. Since the majority of research carried out to date has concentrated on the genetic characterization of *Lactococcus*, much of this chapter will focus on the progress achieved with this particular genus.

9.2 Improvement of dairy starters

Despite much progress achieved in starter technology, there remains considerable room for improvement of the strains currently used by the dairy industry. Infection by phage remains a significant cause of starter failure resulting in substantial economic losses. However, a reliance by industry on phage-insensitive strains in a defined strain approach has meant that many other properties associated with starters, such as their role in flavour development, may have been underscored. The result of this is that a product currently manufactured with defined strains may differ considerably to that which was made previously with mixed strain starters.

In addition, there are considerable opportunities for exploitation of specialized cultures with desired functionalities. These include the incorporation of probiotic cultures proposed to have positive effects on human health and nutrition (e.g. certain lactobacilli or bifidobacteria). Currently, the worldwide market for probiotics continues to increase in response to demands from a more health-conscious consumer. Another particularly exciting area for exploitation of LAB is their potential use as oral vaccines to deliver antigens at mucosal surfaces (Wells *et al.*, 1996). These organisms provide excellent candidates for such live vectors given their GRAS (generally regarded as safe) status and associated history in food fermentations. Such an approach involves the expression of the foreign antigens on the surface of the vector cell where it can evoke a host immune response.

It may also be desirable for starter bacteria to produce antimicrobial substances such as bacteriocins in dairy fermentations. Bacteriocins are peptides or small proteins produced by some lactic acid bacteria which display either narrow or broad spectrum antagonistic activity against other Gram positive bacteria. These antimicrobials can provide an extra hurdle against pathogenic or spoilage organisms in a fermented foodstuff. For example, starters producing bacteriocins such as nisin or lacticin have been used to improve the microbial quality of a variety of fermented products including Cheddar cheese (Roberts *et al.*, 1992; Ryan *et al.*, 1996).

Specialized starters which overproduce certain flavour compounds would also be a welcome addition to the range of starters currently available to industry. An example of the application of such cultures is the use of a starter which produces large amounts of the flavour compound diacetyl used in the commercial production of lactic butter by the Dutch NIZO process (Hugenholz, 1993). Such strains could also be used as starter adjuncts in combination with fast-acid 'reliable' starters, thus ensuring that adequate acid development takes place in combination with the additional positive attributes afforded by the adjunct culture. Genetic approaches could be used to convert traditional starter strains to ones possessing these novel activites.

9.3 Genetic tools available in lactococci

Larry McKay of the University of Minnesota was the first researcher to demonstrate that the lactococci contain plasmids (McKay and Baldwin, 1975). These extrachromosomal elements were initially visualized under the electron microscope, but were soon confirmed by more conventional electrophoresis techniques. With hindsight, the presence of plasmids was predictable in that it had long been noted that many of the key industrial traits exhibited by the lactococci were notoriously unstable, and were frequently lost. Most lactococcal strains possess multiple copies of between four and seven distinct plasmids, which have been shown to encode many of the traits essential for efficient fermentations. In addition, it was rapidly determined that a number of these plasmids were conjugative, and could be transferred between strains at low frequencies. Investigators were quick to recognize the advantages of the presence of a plethora of native plasmids in these industrially important strains, and were eager to exploit this useful biological phenomenon.

9.3.1 Assembling the tools

The basic requirements of a good genetic system demand a suitable plasmid-free host, a versatile vector system and a means of introducing recombinant DNA into cells. At the beginning of the 1980s none of these essential requirements was available, but, due to the intensive efforts of a number of key research groups, by the mid 1980s all of the following features were in place.

9.3.1.1 Plasmid-free hosts. One of the seminal papers in the exploitation of lactococci as potential hosts for recombinant DNA research followed the observation that successive rounds of protoplast regeneration can lead to the formation of a plasmid-free derivative of an industrial strain (Gasson, 1983). This plasmid-free background was ideal for studying native lactococcal plasmids introduced by conjugation (a term explained in section 9.4). This was also a suitable strain to act as a host for recombinant DNA plasmids.

9.3.1.2 Suitable vectors. The second requirement of a functional genetic system is the availability of useful vector systems with a range of biological functions. Central to the creation of suitable plasmid vectors was the observation that pWV01, a small cryptic plasmid from *Lactococcus lactis* subsp. *cremoris* Wg2, was able to replicate in *Bacillus subtilis* and *Escherichia coli* (Kok *et al.*, 1984). This plasmid and its derivatives carrying suitable features such as selectable antibiotic resistance markers and multiple cloning sites were the first lactococcal cloning vectors, and were soon followed by other plasmids based on additional small cryptic plasmids. The construction of versatile second generation plasmid vectors based on pWV01 soon followed (see below).

9.3.1.3 Transformation. The final requirement of transformation protocols was achieved through the use of protoplast formation and regeneration in the presence of naked DNA (based on systems initially developed in *B. subtilis* and *Streptomyces*). Protoplast transformation of lactococci was first described in 1984, but the frequency was extremely low (Kondo and McKay, 1982). This cumbersome and unreliable technique represented a major breakthrough in the analysis of lactococcal gene systems, but it has been largely replaced by the more reliable technique of electroporation (electrotransformation) which became widely available at the end of the decade.

Thus, the development of transformation protocols, the availability of suitable plasmid free strains such as MG1363, and the versatility of the pWV01 derived vectors opened the way for the genetic analysis of genes and gene systems in LAB.

9.3.2 *The development of more sophisticated techniques*

The original cloning vectors based on the cryptic plasmids pWV01 and pSH71 were little more than a simple replicon containing a selectable marker (usually either erythromycin or chloramphenicol resistance) and a multiple cloning site. While these simple vectors were the workhorses of much of the early analysis of the lactococci, they have been refined and adapted for a number of different tasks in recent years. Perhaps the earliest example of a second generation vector was pGKV210, a promoter screening vector consisting of the pWV01 origin, a selectable marker (Ery) and a multiple cloning site located immediately upstream of a promoterless Cm gene (van der Vossen *et al.*, 1985). Random fragments were cloned in this multiple cloning site and the recombinant plasmids were screened for Cm (chloramphenicol) resistance. Variations on this basic scheme included terminator screening vectors, and a series of expression plasmids based on the promoters originally isolated using pGKV210.

Perhaps the next most important development was that of plasmid integration initially reported by Leenhouts from the Venema group (1991). The early integration vectors could not replicate in the lactococcal background. Thus, any homology between the DNA cloned in the integration plasmid and the genome could result in homologous recombination between the plasmid and the chromosome. In a more recent development the replication gene is borne on a temperature-sensitive replicon while the integration vector bears the replication origin. In this system, the integration plasmid can be introduced at the permissive temperature and allowed to establish itself within the host. At the shift to a higher temperature, the recently introduced plasmid is forced to integrate. This system helps to overcome the low frequency at which homologous recombination normally occurs in lactococci. Whichever integration system is employed, the net result is the ability to introduce genes in a stable single copy, rather than be restricted to reliance on plasmid-borne determinants.

Another approach which is the subject of continued interest is the prospect of self-cloning to allow the construction of recombinant plasmids which will be permitted in food grade strains. The basic principle of self-cloning is that DNA of solely lactococcal origin is used to construct smaller replicons from existing large plasmids. For example, a large lactococcal plasmid encoding phage resistance, or bacteriocin production could be whittled down to a smaller plasmid bearing only the information for the trait of interest and the minimal information for the maintenance of the plasmid (replication functions). Ideally, this plasmid downsizing is achieved using detailed sequence information such that the precise consequences of the plasmid rearrangement can be predicted. A logical extension would be the construction of recombinant plasmids derived from a number of different lactococcal replicons and chromosomal genes, but based again on the complete DNA sequence of the relevant fragments.

In order to fully capitalize on advances in our ability to manipulate the genetic complement of the lactococci, it is imperative that recombinant molecules can be introduced in a food grade fashion. This requires, in the first instance, that no antibiotic resistance markers are employed, but also that the effects of any such manipulations are fully understood. This criterion can only be fully realized if a genetic system is used in which literally every nucleotide is known, and ideally in a situation in which no DNA other than that of lactococcal origin is introduced to the strain destined for the food chain. This is not an unreasonable demand, since a number of the small cryptic plasmids of the lactococci have been fully sequenced, and in some instances the genes which one might like to introduce to the starter strain may be selectable in themselves. For example, it is not difficult to imagine an instance in which a fully sequenced fragment bearing a bacteriophage resistance gene could be introduced into a sequenced cryptic plasmid and transformed into an industrial strain. Successful transformants could be selected on their ability to withstand bacteriophage attack. Equally, a bacteriocin production and immunity operon could be introduced in a similar fashion. Such strains could be regarded as food grade, in that no DNA other than that of lactococcal origin is employed, and the combined sequence could be analysed to ensure that no inadvertent genetic consequences result from the hybrid plasmid (i.e. switching on of previously silent open reading frames). In those instances in which the gene of interest does not have a readily selectable phenotype it is necessary to incorporate a food grade selectable marker into the cloning vehicle. Once again, bacteriocin immunity or bacteriophage resistance are two potential markers. In addition systems have been designed based on the *lac* genes which allow either for selection, or insertional inactivation of the lactose phenotype (Platteeuw *et al.*, 1996).

It is also tempting to utilize a system in which a genetically manipulated gene is reintroduced via replacement recombination to its correct chromosomal location. In such a fashion quite specific alterations to the lactococcal genome could be targeted in a very precise manner. There would be difficulties with this system to define suitable selection conditions, but this should not prove to be an unsurmountable problem.

9.4 Genetic transfer in lactic acid bacteria

The ability to introduce genetic material into and exchange it between bacterial hosts is prerequisite to the genetic analysis and manipulation of any bacterial system. Three main avenues are available for the introduction of

DNA to bacterial cells: transduction, conjugation, and transformation. All three gene exchange mechanisms have been demonstrated in the lactococci (reviewed by Gasson and Fitzgerald, 1994).

9.4.1 Gene exchange mechanisms

9.4.1.1 Transduction. This refers to gene transfer mediated by phage. In the lactococci, transfer of the lactose phenotype between closely related strains was accompanied by 'transductional shortening' of the 45 kb plasmid to a 30 kb molecule. In fact, this shortening most probably reflects the size selection of the bacteriophage capsid for spontaneously deleted plasmids, rather than any deliberate shortening event during packaging. This smaller plasmid can be transduced at much higher frequencies in subsequent transduction events. Despite these high frequencies, transduction has not been employed to any great extent in the genetic manipulation of the lactococci.

9.4.1.2 Conjugation. The natural conjugative ability of many lactococcal plasmids has provided researchers with a convenient means of introducing genetic material to target strains in a 'natural' non-recombinant fashion. The process relies upon the transfer of genetic material as a consequence of physical contact between donor and recipient. Conjugation frequencies as high as 10^{-1} per recipient have been recorded, but frequencies of between 10^{-5} to 10^{-7} per recipient are more typical. Conjugative plasmids are commonplace in lactococci, and in some instances these conjugal elements can promote transfer of otherwise non-conjugative plasmids. The ability of native plasmids to conjugatively transfer between strains has formed the basis of a number of strain improvement programmes (section 9.6).

9.4.1.3 Transformation. No natural transformation has been reported for members of the LAB. This is a significant stumbling block in that it limits the use of the most convenient method of introducing novel genetic material at the laboratory bench. Some debate surrounds the 'acceptability' of transformation (whether mediated by protoplast or electro-transformation) as a means of creating strains destined for the food chain. In the European Union, such methods are deemed as non-natural, and are therefore not used to construct strains, other than for the purposes of analysis of key genetic systems. In the USA, the situation appears to be more relaxed, although some researchers are reluctant to use electroporation for strains destined for human consumption.

9.5 Cloning and characterization of key lactococcal traits

The plasmid location of many of the industrially important genes made them attractive candidates for initial genetic analysis and provided the impetus for a major research effort to characterize and manipulate LAB genes. In a productive five-year period, the genes for lactose transport and catabolism, casein utilization, citrate transport, and many different genes encoding bacteriophage resistance, and bacteriocin production and immunity have all been cloned, sequenced and expressed in lactococcal hosts. In addition to these homologous genes and gene systems, a number of heterologous genes have also been successfully expressed in lactococcal hosts. In recent years attention has begun to focus on the chromosome (Davidson *et al.*, 1996) and bacteriophage genomes in addition to plasmid-encoded traits. Rather than provide an exhaustive review of each of these achievements, we will attempt to illustrate the progress made in lactococcal genetics with a few key examples.

9.5.1 *Casein degradation*

Perhaps the earliest complete study of a lactococcal gene system concerned the lactococcal proteinase (reviewed by Kok, 1990). The ability to utilize the milk protein casein is a vital, though somewhat unstable, characteristic of strains used in cheesemaking. At least two proteinase enzymes can be distinguished on the basis of their degradation of casein, termed PI and PIII. Plasmid curing experiments confirmed a plasmid location for these proteinase determinants in strain Wg2 (PI) and strain SK11 (PIII). The relevant genes were cloned in pWV01-derived vectors and expressed in a plasmid-free host. Subsequently both genes were sequenced and compared to one another. Both proteinases are produced as pre-pro-enzymes encoded by a single gene of approximately 6-kb, designated prtP. The different specificities exhibited by the enzymes can be ascribed to a small number of amino acid substitutions. This work has been extended recently to elegantly replace regions from one gene with the relevant region of the other, leading to the production of novel proteinases with unique casein specificities. A second significant finding was that a second gene, prtM, is required for the maturation of the proteinase. This second gene is located immediately adjacent to the prtP gene on both plasmids. Subsequent analyses of proteinase genes encoded by other lactococci have confirmed this two-gene arrangement as a common feature of these industrially important systems.

9.5.2 *Bacteriophage resistance*

Bacteriophage remain the single most significant cause of starter failure in the dairy industry. As a consequence, the ability to withstand bacteriophage attack is an extremely important feature of cheesemaking strains. Lactococci exhibit at least four distinct mechanisms for resisting bacteriophage attack: adsorption inhibition (Ads), injection blocking (Inj), abortive infection (Abi) and restriction/modification (R/M) (reviewed by Hill, 1993 and Garvey *et al.*, 1995a). There are a number of independent examples of each mechanism, and each has an associated genetic determinant(s). Early studies revealed that many of these mechanisms are plasmid-encoded, and very quickly phenotypes were linked to plasmid species. In some instances these native lactococcal plasmids are conjugative, and this has been exploited to good effect by the dairy industry to construct long-lived resistant strains using the natural process of conjugation (section 9.6). A number of these phage resistance genes have been cloned and sequenced using classical shotgun cloning and expression studies.

9.6 'Natural' genetic modification of lactococcal starters

The application of traditional approaches to starter strain improvement has yielded limited success. One of the best examples is the frequent use and isolation of bacteriophage insensitive mutants (BIMs). BIMs can be generated for most strains simply through exposure to phage or phage-containing factory wheys. Such variants normally arise as a consequence of spontaneous point mutations occurring on the host chromosome which results in an inability of the phage to adsorb to the host. In one case, however, the isolation of a BIM was associated in the co-integration of two resident plasmids in a lactococcal strain (Harrington and Hill, 1992). Interestingly, the resulting BIM had acquired a new R/M (restriction/modification) and adsorption activities, presumably associated with the formation of the co-integrate plasmid. The exploitation of BIMs provides a very rational and straightforward approach to starter strain improvement and in many instances has proved very successful in commercial practice. There are a number of disadvantages associated with the BIM approach, however, which limit its long-term practical usefulness for the generation of resistant strains. These include the inability of some mutants to grow well in milk or a tendency to revert to phage sensitivity. Moreover, it has proved impossible to isolate BIMs for some starter strains. Consequently much research effort in recent years has been in the development of directed food grade genetic approaches to improve starter cultures.

Much of the early pioneering work in lactococcal genetic research concentrated on plasmids and plasmid-encoded functions. As mentioned earlier (section 9.3), the lactococci possess a large and diverse plasmid complement which encodes a number of industrially relevant traits including genes for lactose catabolism, proteinase production and bacteriocin production and immunity. The association of such traits with conjugative plasmids has proved extremely fortuitous since it has facilitated the transfer of commercially important traits between strains (Sanders *et al.*, 1986). A prerequisite to the transfer of naturally occurring plasmids, however, is that they possess a suitable selectable marker that can be used for identification of transconjugants. Plasmid-linked phage resistance determinants have been used in numerous conjugations to date, with transconjugants being selected for on the basis of improved resistance to a homologous phage (Figure 9.2). This approach was first adopted by Klaenhammer and Sanozky (1985) for transfer of pTR2030 from a *Lac*-donor to an industrial starter. Many of the transconjugants produced in this way have performed admirably under commercial production conditions, in a situation in which the parental strains would have rapidly succumbed to phage attack. In a similar fashion, Jarvis *et al.* (1989) used this method to transfer the plasmid pAJ1106 to a number of strains.

When using phage for selection, it is necessary that the phage resistance plasmid being transferred should provide adequate resistance, as partially protected strains would be less likely to survive exposure to phage in the selection process. In addition, bacteriophage insensitive mutants (BIMs), which are phenotypically indistinguishable from putative transconjugants, can develop at high frequencies during this type of selection. Thus it can be advantageous if the conjugal plasmid contains additional selective markers such as resistance to bacteriocins. Using both phage and nisin resistance, properties associated with the plasmid pNP40 were transferred to a strain (DPC220) used in the commercial production of lactic butter (Figure 9.3) (Harrington and Hill, 1991). DPC220 is a component of a mixed strain starter but is indispensable since it overproduces the diacetyl necessary for the flavour of the product. The use of DPC220 is fraught with difficulty since it has proven to be unreliable under prolonged industrial use due to its inherent phage sensitivity. The resultant transconjugant containing pNP40

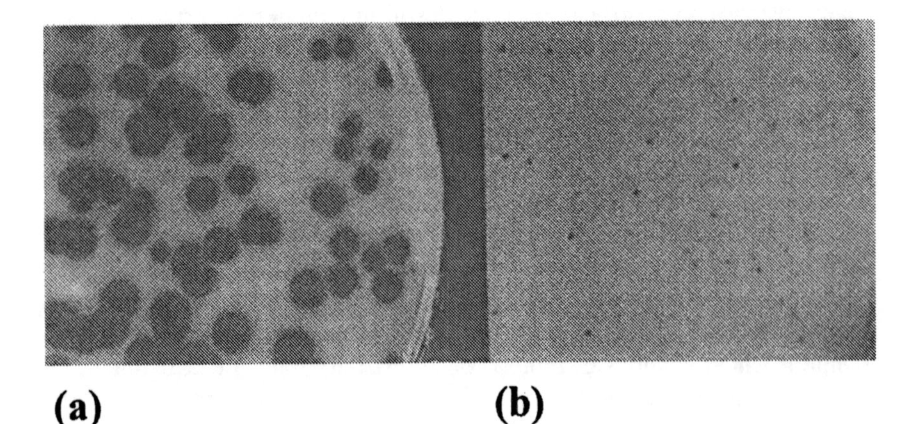

Fig. 9.2 The conjugal transfer of phage resistance plasmids to lactococcal starters renders them less sensitive to phage attack. In this case, the plaque size for a virulent lytic phage is greatly reduced with a transconjugant strain which has received the plasmid pMRC01 (b). In contrast, the bacteriophage forms very large plaques in the parent starter strain (a).

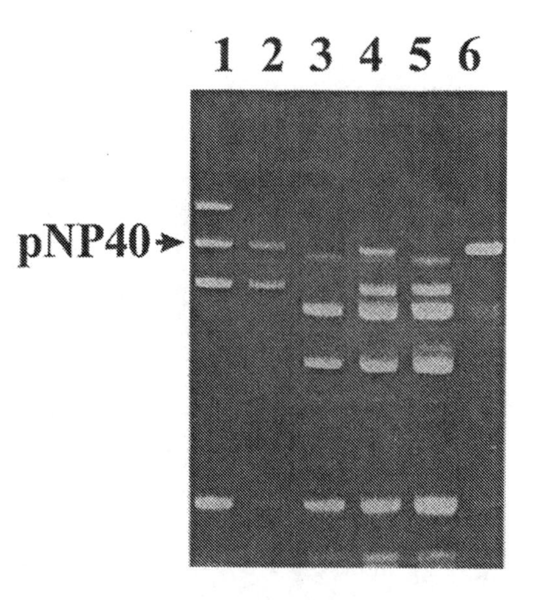

Fig. 9.3 Agarose gel electrophoresis demonstrating the transfer of the phage resistance plasmid pNP40 from a lactose deficient donor strain (lane 2) to a *Lactococcus lactis* subsp. *lactis* biovar *diacetylactis* starter strain used for lactic butter manufacture (lane 3). The resultant transconjugant (lane 4) exhibited excellent phage resistance and was subsequently introduced into Irish industry.

produced similar quantities of diacetyl as the parent but was resistant to attack by homologous phage. This plasmid-containing strain has been subsequently used to replace the existing strain in the Irish industry with considerable success.

The presence of lactose fermenting determinants on a plasmid harbouring phage resistance genes (e.g. pTR1040) offers an alternative naturally-occurring selectable marker. In such a system, a lactose deficient derivative of the cheese-making recipient has to be generated prior to introduction of such a phage resistance plasmid. However, interference with the natural lactose system of industrial starters is not favoured, diminishing the usefulness of this approach for strain improvement.

Another example of the manipulation of lactococcal starters involves the conjugal transfer of bacteriocin production and immunity plasmids to starter strains in natural conjugations. For example, pMRC01, a conjugative plasmid conferring the ability to produce the broad host range bacteriocin lacticin 3147, has been successfully transferred to a number of dairy starters (Ryan *et al.*, 1996) (Figure 9.4). These transconjugants are capable of repressing the appearance of non-starter LAB during cheese ripening.

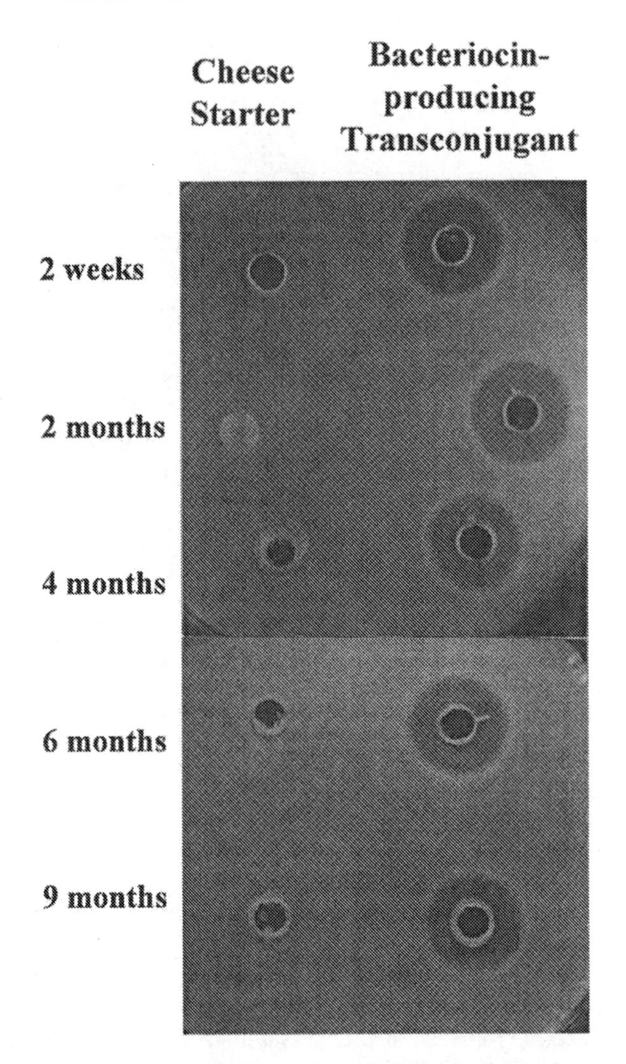

Fig. 9.4 The transfer of genes responsible for bacteriocin production to cheese starters can be an efficient approach for incorporating these natural antimicrobials into food products. This figure shows the residual lacticin 3147 (as indicated by inhibition zones on seeded agar plates) in cheese made with a commercial starter and its bacteriocin producing transconjugant derivative over a nine-month ripening period.

9.7 Metabolic engineering of lactococci

Metabolic engineering refers to the redirecting of metabolism such that the flux towards wasteful or energy-utilizing steps is reduced or eliminated. Using such an approach, lactococci can be manipulated to overproduce certain valuable metabolic products which are normally generated at low levels, or to produce new metabolites not normally made by these strains (de Vos, 1996). Since lactococci have multiple applications in dairy fermentations, such end-products could potentially be incorporated directly into food products by such an approach. Lactococci represent excellent candidates for metabolic engineering experiments given their GRAS status and limited catabolic flexibility. Their strictly fermentative nature is due in part to the absence of a functional electron transport chain. In addition, the gradual adaptation of many dairy strains to growth in milk has meant that they can use only mono- and di-saccharides as energy sources. Many of these dairy strains also lack the ability to synthesize certain amino acids, for example, branched chain amino acids (Godon *et al.*, 1993) and histidine (Delorme *et al.*, 1993) in contrast to lactococci isolated from non-dairy sources which are generally prototrophic.

A prerequisite to efficient metabolic engineering in any strain is a comprehensive understanding of the physiology of the organism concerned. In many cases, dramatic changes in the relative amounts of end products synthesized can be achieved simply by manipulating environmental parameters. Such approaches, however, have their limitations in that strains cannot be altered to overproduce certain products if they lack the essential genetic determinants which enable them to synthesize the product in the first place. For this reason, a combined physiological/genetic approach has tremendous promise to expand the metabolic potential of organisms such as lactococci. The feasibility of adopting such an approach was recently demonstrated by Hugenholtz *et al.* (1996) who engineered *L. lactis* to produce ethanol.

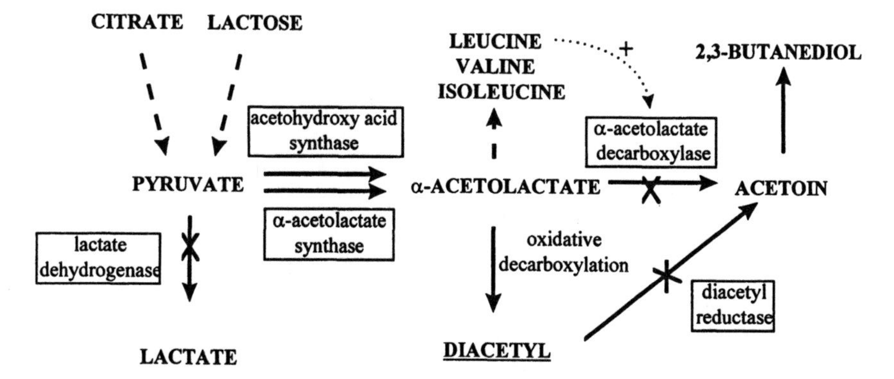

Fig. 9.5 Metabolic engineering strategies for overproducing diacetyl in lactococci target both the overproduction of acetolactate and block the formation of acetoin. Overproduction of acetolactate synthase or overproduction of acetohydroxy acid synthase in combination with inactivation of lactate dehydrogenase have caused acetolactate to be produced in far higher quantities. In addition, inactivation of acetolactate decarboxylase blocks the formation of acetoin from acetolactate which causes diacetyl to be overproduced.

Considerable effort has also focused recently on metabolic engineering approaches to increase the level of diacetyl produced by *L. lactis* subsp. *lactis* biovar *diacetylactis*. Diacetyl, an important flavour compound, is produced from citrate which is co-metabolized with lactose (Hugenholz, 1993). The metabolic precursor of diacetyl is acetolactate which is, in turn, generated from pyruvate either by the enzyme, acetolactate synthase, or acetohydroxy acid synthase (Figure 9.5). The acetolactate is then either converted to acetoin by the enzyme, acetolactate decarboxylase, or is non-enzymatically decarboxylated to diacetyl in the presence of oxygen. The cloning and over-expression of the genes for the enzymes implicated in production of diacetyl has led to an increase in product formation (Platteeuw *et al.*, 1995; Benson *et al.*, 1996). Similarly, it has been observed that deletion of the genetic determinant for acetolactate decarboxylase, the enzyme which converts acetolactate to acetoin, resulted in production of significantly higher amounts of diacetyl (Swindell *et al.*, 1996; Goupil *et al.*, 1996).

9.8 Social, cultural and legal aspects and implications

LAB constitute the vast majority of organisms used worldwide in dairy fermentations. The resultant food products, most of which contain live organisms (unless pasteurized), are perceived by consumers as safe, healthy and nutritious. It is obviously of the utmost importance that the acceptance of these products by consumers is not compromised in any way as a result of the application of molecular biology techniques. In this regard, consumers need to be informed that altered strains are safe and that they provide a distinct benefit to the product over and above that afforded by the parent organisms (for a recent review, see Verrips and van den Berg, 1996).

A good example of acceptance of genetic approaches in the dairy industry has been the widespread adoption of recombinant chymosin as a substitute for animal-derived rennet in cheesemaking (see Chapter 6 for further details). Although there is a distinction to be made between the addition of a product of a genetically modified organism (i.e. an enzyme) to a food *versus* the addition of a whole genetically modified organism to a food which may be consumed without further processing, it can be anticipated that major improvements in starter performance afforded through genetic approaches would also be welcomed by the dairy industry.

The application of genetically engineered LAB in the European Community is regulated by either of two EC Directives (European Commission Directives 94/15/EC, 1994; 94/51/EC, 1994) and from May 1997 by the Novel Foods Regulation (for more details, see Chapter 4). These regulations define a genetically modified organism (GMO) as 'an organism in which the genetic material has been altered in a way that does not occur naturally by mating or natural recombination'. At present, LAB have been altered for consumer products by natural bacterial conjugation only, facilitating transfer of naturally-occurring plasmids. The resultant transconjugant strains are therefore not considered genetically modified microorganisms and consequently do not fall under current regulations in Europe and can be used for food applications similarly to existing starters. Indeed, the examples of the benefits obtained in starter technology referred to in section 9.6 fall into this category. Electroporation is also accepted in some countries, such as the USA, as an alternative to conjugation for delivery of plasmids into strains.

The fact that most natural plasmids either lack a suitable food-grade selectable marker, are non-transmissible or incompatible with resident plasmids or are lost from the recipient strain at a high frequency, gives rise to limitations which narrow the scope of what can be achieved in starter development through genetic approaches. In addition, it is expected that only a small proportion of plasmids found in LAB would afford significant improvements when transferred to starter strains. Consequently, there is now an increasing interest in the development of food-grade approaches for strain improvement. Where self-cloning approaches are employed (in which all the genetic information in a recombinant plasmid is derived from the host strain), the resultant modified strains are considered safe since the genetic modifications performed in the strains do not involve the introduction of foreign (non-self) DNA. Therefore, it is highly unlikely that any combination of genes from the cheesemaking lactococci could be constructed which would be harmful. A different situation arises where DNA from one food-grade organism is used to improve another food grade organism. In such cases, it can also be argued that the resultant strain is food-grade; nevertheless, such modifications will obviously have to be examined closely on a case-by-case basis.

In an effort to assess whether modified DNA could be transferred from lactococci to other bacteria, studies have been undertaken to investigate the genetic transfer frequencies in fermentors, cheese and the digestive tract of mice (Economides, 1991). These have demonstrated that under optimum laboratory conditions, transfer of genes on self-transmissible plasmids occurs at high frequencies between lactococcal strains but that transfer of genes on non-transmissible plasmids occurs at low frequencies and only when a self-transmissible plasmid is also present. In contrast, transfer of chromosomally encoded genes was not detected. In fermentors and in cheese, genetic transfer efficiencies were far lower and could be detected only for genes harboured by self-transmissible plasmids. With regard to the digestive tract, lactococci did not become established for a significant length of time. These studies thus indicate that the possibility of genetic exchange from genetically engineered lactococci should not pose a health risk; however, this is an area which obviously requires more intensive investigation especially where LAB harbour foreign (non-self) DNA.

Explicit demonstration of the benefits of LAB which have been manipulated by natural means (as described in section 9.6) could be regarded as a prerequisite for the widespread consumer acceptance of recombinant DNA approaches in the dairy industry. An example where such benefits are obvious is the manipulation of starters to produce broad spectrum bacteriocins such as nisin (Steele and McKay, 1986) or lacticin (Ryan *et al.*, 1996). Since these bacteriocins inhibit a wide range of spoilage and pathogenic bacteria, foods made with such starters would exhibit major improvements with regard to both their quality and safety (Stiles, 1996). In addition, genetic strategies used to enhance the probiotic potential of LAB would be welcomed by the modern health-conscious consumer. Such studies include their exploitation as oral vaccines which could be administered in the form of a fermented dairy product. Improvements in the production and efficiency of manufacture of fermented foods as a result of using genetically modified LAB should also result in concomitant reductions in the cost of such products. Thus, it is very likely that the safety and potential benefits to be gained from the use of genetically modified LAB will be fully realized in the future which would ensure their adoption for the manipulation of starter cultures for industry.

9.9 Conclusions

A great deal of research effort has been directed towards the genetic analysis of the lactococci in the last decade. The industrial benefits of this research are largely unrealized, not because of technological limitations, but rather as a consequence of a reluctance to introduce recombinant strains into foods. These reservations have been largely overcome in the case of natural transconjugants exhibiting increased phage resistance, as outlined in section 9.6. The future exploitation of molecular techniques available in the lactococci will largely depend on attitudes, both of regulators and consumers. It is probable that the acceptance of recombinant cultures will require a responsible attitude on the part of the scientist and industrialist working in tandem through education of the consumer. It must be accepted, however, that the final verdict on the acceptability or otherwise of genetically altered starter bacteria in dairy products will rest with the consumer. This realization places a heavy burden on researchers and product developers to take account of consumer concerns, and not to be dismissive of consumer fears. In this chapter, we have attempted to illustrate the advances made in recent years in the genetic manipulation of starter organisms, but we acknowledge that many of these advances remain on the laboratory bench and have not progressed to the supermarket shelf. A bright future awaits the LAB, if researchers and consumers (which are not mutually exclusive groups) can demonstrate the significant benefits of using constructed strains in economic terms, and not simply as a reflection of the abilities of researchers to perform certain manipulations.

References

Benson, K. K. *et al.* (1996) Effect of *ilvBN*-encoded acetolactate synthase expression on diacetyl production in *Lactococcus lactis*. *Appl. Microbiol. Biotechnol.* 45, 107–11.

Commission Directive 94/51/EC of 7 November 1994 adapting to technical progress Council Directive 90/219/EEC on the contained use of genetically modified micro-organisms. *Official Journal of the European Communities*, L297, 18/11/94, p. 29.

Commission Directive 94/51/EC of 15 April 1994 adapting to technical progress for the first time Council Directive 90/220/EEC on the deliberate release into the environment of genetically modified micro-organisms. *Official Journal of the European Communities*, L103, 22/4/94, p. 20.

Davidson, B. E., Kordias, N. *et al.* (1996) Genomic organisation of lactic acid bacteria. *Antonie van Leeuwenhoek* 70, 161–83.

De Vos, W.M . (1996) Metabolic engineering of sugar metabolism in lactic acid bacteria. *Antonie van Leeuwenhoek* 70, 223–42.

Economides, I. (ed.). (1991) Biotechnology R and D in the EC. 28–9.

Froseth, B. R. and McKay, L. L. (1991) Molecular characterisation of the nisin resistance region of *Lactococcus lactis* subsp. *lactis* biovar. *diacetylactis* DRC3. *Appl. Environ. Microbiol.* 57, 804–11.

Garvey, P., van Sinderen, D., *et al.* (1995) Molecular genetics of bacteriophage and natural phage defence systems in the genus *Lactococcus*. *Int. Dairy Journal.* 5, 905–47.

Gasson, M. J. (1983) Plasmid complements of *Streptococcus lactis* NCDO 712 and other lactic streptococci after protoplast-induced curing. *J. Bacteriol.* 154, 1–9.

Gasson, M. J. and Fitzgerald, G. F. (1994) Gene transfer systems and transposition. In *Genetics and Biotechnology of Lactic Acid Bacteria*, M. J. Gasson and W. M. de Vos (eds), Chapman & Hall, London, pp. 1–51.

Godon, J.-J., Delorme, C., *et al.* (1993) Gene inactivation in *Lactococcus lactis*: branched chain amino acid biosynthesis. *J. Bacteriol.* 175, 4383–90.

Goupil, N., Godon, J.-J., *et al.* (1996) Imbalance of leucine flux in *Lactococcus lactis* and its use for the isolation of diacetyl-overproducing strains. *Appl. Environ. Microbiol.* 62, 2636–40.

Harrington, A. and Hill, C. (1991) Construction of a bacteriophage-resistant derivative of *Lactococcus lactis* subsp. *lactis* 425A by using the conjugal plasmid pNP40. *Appl. Environ. Microbiol.* 57, 3405–9.

Hill, C. (1993) Bacteriophage and bacteriophage resistance in lactic acid bacteria. *FEMS Microbiol. Rev.* 12, 87–8.

Hugenholtz, J. (1993) Citrate metabolism in lactic acid bacteria. *FEMS Microbiol. Rev.* 12, 165–78.

Jarvis, A. W., Heap, H. A. *et al.* (1989) Resistance against industrial bacteriophage conferred on lactococci by plasmid pAJ1106 and related plasmids. *Appl. Environ. Microbiol.* 55: 1537–43.

Klaenhammer, T. R., and Sanozsky, R. B. (1985) Conjugal transfer from *Streptococcus lactis* ME2 of plasmids encoding phage resistance, nisin resistance and lactose-fermenting ability: evidence for a high frequency conjugal plasmid responsible for abortive infection of virulent bacteriophage. *J. Gen. Microbiol.* 131, 1531–41.

Kok, J., Van der Vossen, J. M. B. M. and Venema, G. (1984) Construction of plasmid cloning vectors for lactic streptococci which also replicate in *Bacillus subtilis* and *Escherichia coli*. *Appl. Environ. Microbiol.* 48, 726–31.

Kok, J. (1990) Genetics of the proteolytic system of lactic acid bacteria. *FEMS Microbiol. Rev.* 87, 15–42.

Kondo, J. K. and McKay, L. L. (1982) Transformation of *Streptococcus lactis* protoplasts by plasmid DNA. *Appl. Environ. Microbiol.* 43, 1213–15.

Leenhouts, K. J., Kok, J. and Venema, G. (1991) Lactococcal plasmid pWV01 as an integration vector for lactococci. *Appl. Environ. Microbiol.* 57, 2562–67.

McKay, L. L. and Baldwin K. A. (1975). Plasmid distribution and evidence for a proteinase plasmid in *Streptococcus lactis*. *Appl. Microbiol.* 29, 546–8.

Platteeuw, C. *et al.* (1995) Metabolic engineering of *Lactococcus lactis*: influence of the overproduction of α-acetolactate synthase in strains deficient in lactate dehydrogenase as a function of culture conditions. *Appl. Environ. Microbiol.* 61, 3967–71.

Platteeuw, C., van Alen-Boerrigter, I. J. *et al.* (1996) Food-grade cloning and expression system for *Lactococcus lactis*. *Appl. Environ. Microbiol.* 62, 1008–13.

Pot, B., Ludwig, W. *et al.* (1994) Taxonomy of lactic acid bacteria in *Bacteriocins of Lactic Acid Bacteria* (eds. L. de Vuyst and E. J. van Damme) Blackie Academic & Professional, Glasgow, pp. 13–90.

Roberts, R. F., Zottola, E. A. *et al.* (1992) Use of a nisin-producing starter culture suitable for Cheddar cheese manufacture. *J. Dairy Sci.* 75, 2353–63.

Ryan, M. P. *et al.* (1996) An application in cheddar cheese manufacture for a strain of *Lactococcus lactis* producing a novel broad-spectrum bacteriocin, Lacticin 3147. *Appl. Environ. Microbiol.* 62, 612–19.

Sanders, M. E. *et al.* (1986) Conjugal strategy for construction of fast acid-producing, bacteriophage-resistant lactic streptococci for use in dairy fermentations. *Appl. Environ. Microbiol.* 52, 1001–7.

Steele, J. L. and McKay L. L. (1986) Partial characterization of the genetic basis for sucrose metabolism and nisin production in *Streptococcus lactis*. *Appl. Environ. Microbiol.* 51, 57–64.

Steenson, L. R. and Klaenhammer, T. R. (1985) *Streptococcus cremoris* M12R transconjugants 331–45.

Swindell, S. R., Benson, K. H. *et al.* (1996) Genetic manipulation of the pathway for diacetyl metabolism in *Lactococcus lactis*. *Appl. Environ. Microbiol.* 62, 2641–43.

Van der Vossen, J. M. B. M., Kok, J., and Venema, G. (1985) Construction of cloning, promoter-screening and terminator-screening vectors for *Bacillus subtilis* and *Streptococcus lactis*. *Appl. Environ. Microbiol.* 48, 726–31.

Venema, G., Huis in 't Veld *et al.* (eds) Proc. 5th Symp Lactic Acid Bacteria: Genetics Metabolism and Applications. Veldhoven, The Netherlands 8–12 September 1996.

Verrips, C. T. and van den Berg, D. J. C. Barriers to application of genetically modified lactic acid bacteria. *Antonie van Leeuwenhoek* 70, 299–16.

Wells, J.M ., Robinson, K. *et al.* (1996) Lactic acid bacteria as vaccine delivery vehicles. *Antonie van Leeuwenhoek* 70, 317–30.

10 Designer oils: the high oleic acid soybean

ANTHONY J. KINNEY and SUSAN KNOWLTON

10.1 Introduction and historical perspective

Over the past three decades, there has been increasing consumer awareness of the health issues related to diet. Of all the nutrients in the modern diet, none has caused more concern to both consumers and nutritionists than dietary fat (Wiseman, 1992). This concern has arisen from the association between certain types of fat in the diet and a number of human diseases, especially cardiovascular disease.

For edible oil suppliers, and for end-users in the food service and food processing industry, the primary concern is functionality of the oil. Since most edible oils are used for either cooking or for margarines (Figure 10.1), two of the most important of these functional properties are oxidative stability and solid fat content (SFC). Edible oils consist mostly of triacylglycerol molecules with three fatty acids attached to a carbon backbone (Figure 10.2). Most vegetable oils contain triacylglycerols which are rich in polyunsaturated fatty acids. These are highly susceptible to oxidation, as shown in Figure 10.2, which results in the fat becoming rancid, that is, developing offensive odors and flavors.

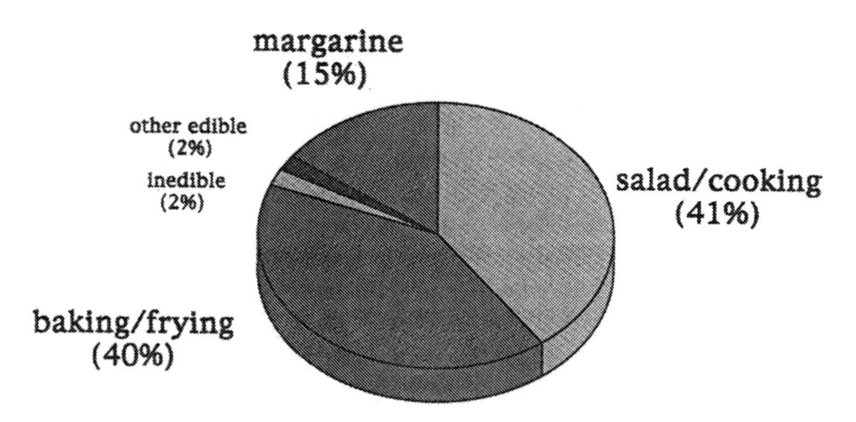

Fig. 10.1 United States market for vegetable oils. The total area of the pie represents more than 15 billion pounds of oil.

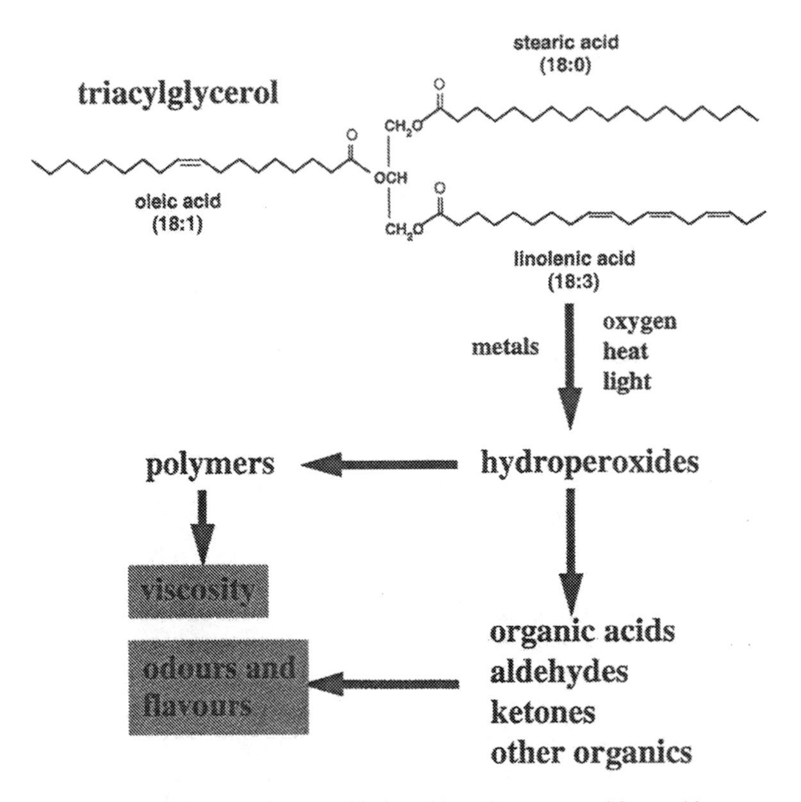

Fig. 10.2 Triacygylcerol molecule and oxidation of its polyunsaturated fatty acid components.

Most edible oils are mixtures of different triacylglycerols which contain different combinations of saturated and unsaturated fatty acids attached to the backbone. Thus, at any given temperature, some of these triacylglycerol molecules will be in a solid phase and some in a liquid phase. The ratio of solid to liquid phases in any oil at a given temperature is expressed as the solid fat content or SFC. A fat with a high SFC at room temperature is hard whereas one with a low SFC is soft or liquid. Also important is the SFC profile over a range of temperatures. Cocoa butters, for example, have a high SFC at room temperature and yet are liquid at body temperature. The shape of the SFC curve between these two temperatures (i.e. the melting profile) determines such important properties as the mouthfeel of chocolate. Therefore, functionality is of primary concern to consumers as well as to industry since the former will not tolerate rancid taste, margarines that are liquid at room temperature, or salad dressings that have an undesirable mouthfeel.

Tropical vegetable oils and animal fats are rich in saturated fats and therefore satisfy both stability and SFC requirements for many applications. However, their use has greatly declined in Europe and the USA because they are perceived as unhealthy. Yet, this decline in the use of animal and tropical oils has only been possible because hydrogenated soybean, rapeseed, cotton and corn oils have been available to replace them.

Hydrogenated vegetable oils are stable because some or all the polyunsaturated fatty acids have been converted to monounsaturated and saturated fatty acids by the chemical addition of hydrogen to their double bonds. The monounsaturated fatty acids in chemically hydrogenated oils, however, are not the same as those found in untreated food fats and oils. The predominant monounsaturated fatty acid in natural oils is *cis*-oleic acid. In some chemically hydrogenated oils, a large percentage of the monounsaturated fatty acid is elaidic acid (Craig-Schmidt, 1992), which is the *trans* isomer of oleic acid. In addition to increasing the oxidative stability of a vegetable oil, elaidic acid also provides some of the SFC functionality in solid fat applications since its melting point (43°C) is more than twice that of oleic acid. Thus it has been possible to produce margarines, for example, that are apparently rich in monounsaturated fatty acids and yet still have the right SFC to be solid at room temperature.

In the past few years the consumption of *trans* fatty acids by humans has also been linked to cardiovascular disease (Ascherio and Willet, 1995) and consumers are becoming more and more aware of the possible consequences of eating hydrogenated oils. In addition, hydrogenated oils often have off-flavours associated with the hydrogenation process (Keppler *et al.*, 1965). The advent of genetically engineered oilseeds has made it possible to produce natural oils that have the desired stability and SFC. These new oils are free from *trans* fatty acids and low in certain saturated fatty acids such as palmitic acid. For the first time, healthiness and functionality need no longer be mutually exclusive properties of an edible oil.

10.2 Target applications in the genetic modification of oil

There have been a number of approaches to produce modified vegetable oils with an SFC similar to that of animal fats or hydrogenated oils containing *trans* fatty acids. For example, researchers at Michigan State University have shown that it is possible, for example, to engineer oilseeds with petroselinic acid in their seed oil (Ohlrogge, 1994). Petroselinic acid is a monounsaturated fatty acid similar to oleic acid except that the position of the double bond is six carbons from the carboxyl end of the molecule rather than nine. Like elaidic acid, its melting point (33°C) is about twice that of oleic acid. Oil containing petroselinic acid was made in tobacco seeds by expressing a delta-6 desaturase from coriander, which normally makes petroselinic acid in its seed oil. The total content of petroselinic acid, however, was low and other technical difficulties have prevented this particular fatty acid from becoming a commercial viability (see Ohlrogge, 1994, for discussion).

Stearic acid is thought to be a benign saturated fatty acid in terms of cardiovascular disease and yet has a melting point of about 70°C. Thus, oils rich in stearic acid would be ideal for providing the SFC required for solid fat applications. DuPont, Calgene (Monsanto) and Inter-Mountain Canola (Cargill) have produced genetically engineered soybean (DuPont) and rapeseed (Calgene, IMC) oils that are rich in stearic acid (Knutzon *et al.*, 1992; Kinney, 1996). In each case, this was achieved by preventing the synthesis of the plants' own delta-9 desaturase enzyme, which is responsible for the conversion of stearic acid to oleic acid in the developing oilseed. These oils are still in an early stage of development and are not expected on the market for a number of years.

The most commercially advanced of the oils with a modified SFC is Calgene's high lauric acid rapeseed oil (Del Vecchio, 1996). This oil was designed as a substitute for food applications currently using palm kernel oil. By expressing the gene encoding a thioesterase enzyme from California Bay Laurel, Calgene have produced a rapeseed with about 40% of the total fatty acids in the seed oil present as lauric acid (a 12-carbon fatty acid). Because of the way the developing rapeseed incorporates 12-carbon fatty acids into triacylglycerol, the modified oil has lauric acid molecules only at the two outer positions on the glycerol backbone. The center fatty acid is always an eighteen carbon unsaturated fatty acid. When the new oil is fully hydrogenated, this center position fatty acid is converted to stearic acid and thus the structure of most of the triacylglycerol molecules in the oil is laurate-stearate-laurate (LSL). The SFC curve for this fully hydrogenated LSL oil is remarkably similar to cocoa butter and thus the new oil can be used as a cocoa butter substitute in confectionery coatings (Del Vecchio, 1996). Calgene are currently evaluating the new oil for use in other high SFC applications such as coffee whiteners, imitation cheeses and spreads.

There are possible nutritional benefits of using high lauric acid rapeseed oil instead of palm kernel oils. Palm kernel has about 50% of its total lauric acid at the centre position (sn-2 position) of the triacylglycerol molecule whereas the fully hydrogenated high lauric acid canola (rapeseed oil) has only stearic acid. When animals, including humans, digest triacylglycerol, the outer two fatty acids are first removed by enzymes in the gut and the fat is then absorbed as a sn2-monoacylglycerol. Unlike sn2-monoacylglycerols containing lauric acid, those containing stearate or unsaturated fatty acids have not been linked to cardiovascular disease.

Oils that are produced for high SFC applications need to be oxidatively stable and thus free from polyunsaturated fatty acids. Partially hydrogenated liquid frying and cooking oils also have to have good oxidative stability and preferably be low in saturated fatty acids, particularly palmitic acid. Thus reducing the polyunsaturated fatty acid content of plant oils is an important functional modification for almost all of the edible oil applications shown in Figure 10.1 above.

The key technology for reducing the polyunsaturate content of seed oils is the ability to silence the expression of another desaturase gene, the delta-12 desaturase. The delta-12 desaturase converts oleic acid to linoleic acid, which in turn is converted by a third desaturase to linolenic acid (Figure 10.3). Thus, blocking the expression of the delta-12 desaturase prevents the formation of most of the polyunsaturated fatty acids by the seed (a small proportion of the total polyunsaturates in the seed are made by other desaturases not discussed here). Using this technology, DuPont have been able to produce an oxidatively stable, high oleic soybean oil which is now in the early stages of commercial development and is discussed in detail in the rest of this chapter. The technology described here is applicable to other oilseeds, such as rapeseed, cotton and corn. It can be combined with other traits, such as the high stearic acid oil described above, to provide both suitable SFC and oxidative stability. This specific example illustrates the potential to produce healthy plant oils with the functionality requirements of the food industry without the need for chemical hydrogenation or costly fractionation.

10.3 Development and composition of high oleic acid soybeans

The synthesis of polyunsaturated fatty acids in developing oilseeds is catalysed by two membrane associated desaturases (Figure 10.3) which sequentially add a second and third double bond to oleic acid (Kinney, 1994). The second double bond is added at the d-12 (n-6) position by a d-12 desaturase, encoded by the GmFad2-1 gene (Heppard *et al.*, 1996). The third double bond is added at the n-3 (d-15) position by a n-3 desaturase, encoded by the GmFad3 gene (Yadav *et al.*, 1993).

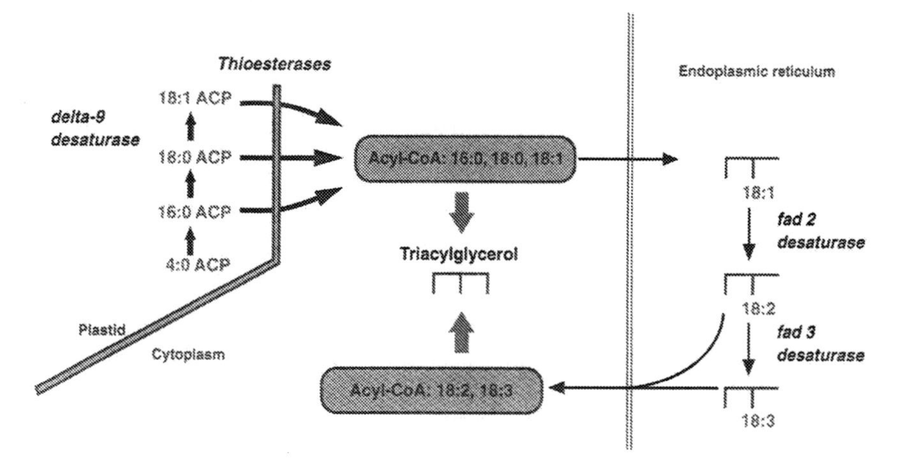

Fig. 10.3 Oil biosynthesis in developing oilseeds.

In soybean (*Glycine max*), there are two Fad 2 genes, one of which (GmFad2-1) is expressed only in the developing seed (Heppard *et al.*, 1996). The expression of this gene increases during the period of oil deposition, starting around 19 days after flowering, and its gene product is responsible for the synthesis of the polyunsaturated fatty acids found in soybean oil. The other gene (GmFad2-2) is expressed in every part of the soy plant at a constant level and its gene product is responsible for the synthesis of polyunsaturated fatty acids for cell membranes.

A transgenic soybean line with an oleic acid content of around 85%, compared with about 24% in conventional soybean oil has been developed. The polyunsaturated fatty acid content of these lines is less than 5%, as shown in Table 10.1. These transgenic soybeans were made by transforming an elite soybean line, Asgrow Elite A2396, with a copy of the soybean GmFad2-1 gene. The transgene was in a 'sense' or forward orientation under the control of a strong, seed-specific promoter derived from the a'-subunit of the *Glycine max* b-conglycinin gene. This promoter allowed high level, seed specific expression of the trait gene (Figure 10.4). The details of the transformation process and molecular analysis of these lines are beyond the scope of this chapter, but can be found elsewhere (Kinney, 1996).

The inserted GmFad2-1 transgene caused a coordinate silencing of itself and the endogenous GmFad2-1 gene. This phenomenon, known as 'sense suppression', is an effective method for deliberately turning off genes in plants and is described in a US patent (Jorgensen and Napoli, 1994). The method has been successfully used by the DNAP Company to produce tomatoes with increased ripening time. Suppression of the GmFad2-1 gene

Table 10.1 Complete fatty acid composition of oil from elite and genetically modified high oleic acid soybean lines grown in the field in 1996

Fatty acid	Elite line (control)	GM high oleic acid line A	GM high oleic acid line B
14:0	trace	trace	trace
16:0	10.1	6.3	6.6
16:1	0.1	0.1	0.2
16:2	trace	trace	trace
16.3	trace	trace	trace
18:0	3.2	3.7	3.6
18:1	14.7	84.6	84.9
18:2 (9, 12)	61.6	0.9	0.6
18:2 (9. 15)	not detected	0.8	0.7
18:3	9.5	2.4	1.9
20:0	0.2	0.4	0.5
20:1	0.2	0.4	0.4
22:0	0.3	0.4	0.5
22:1	trace	trace	trace
24:0	0.1	0.1	0.2

and not the GmFad2-2 gene in soybean resulted in greatly increased oleic acid content in the seed and not in any other tissues of the plant.

A fatty acid has been detected in the high oleic soybean oil which, although common in many edible oils, is not present in the oil of non-transgenic soybeans. This fatty acid was identified by a combination of HPLC and GC-MS as *cis*-9, *cis*-15-octadecadienoic acid which is the 9,15 isomer of linoleic acid. The isomer was probably a product of the activity of a d-15 (n-3) desaturase (GmFad3) which normally inserts a d-15 double bond, into 9,12-linoleic acid. In the transgenic plants, the linoleic acid content was reduced from over 50% of the total fatty acids to less than 2%. Linoleic acid is the normal substrate for the d-15 desaturase. Since this substrate has been greatly reduced in concentration, the d-15 desaturase probably creates a small amount of the isomer by putting a d-15 double bond into 9-oleic acid. This view is supported by the results of crossing the high oleic soybeans with soybean containing a suppressed GmFad3 gene. In the resulting progeny, the isomer was either reduced or eliminated.

The storage protein profile of the genetically modified high oleic soybeans was also found to be different from that of the parent line. In the transgenics, the concentration of b-conglycinin a and a′ subunits was reduced and replaced with glycinin subunits. This was a result of silencing of the a and a′ subunit genes mediated by the a′ promoter sequence used in the GmFad2-1 vector (Kinney, 1996).

10.4 Field performance of high oleic acid soybeans

High oleic acid transgenic soybean lines were planted in the summers of 1995 and 1996 in field plots at a number of different sites in the USA. The

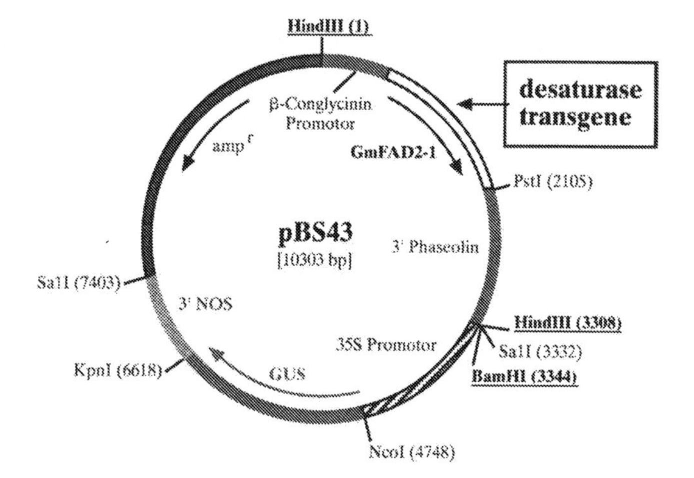

Fig. 10.4 DNA vector used to transform soybeans with the GmFad2-1 gene.

same lines were also grown in the following winters at sites in Puerto Rico and Chile. The bulk oleic acid content of seeds from plants grown at all these sites was about 85% (range 84–88% at each site). The high oleate phenotype was shown to be stable in a number of different environments. By contrast, the oleate content of normal soybeans, and of high oleic acid lines produced by conventional mutagenesis, varied considerably across these different sites.

Yield trials from plants grown in the USA during the Summer of 1995 have indicated that the transgenic plants were indistinguishable from the elite parental lines. More extensive trials in the Summer of 1996 confirmed that the transgenic soybeans were yielding from 50 up to 59 bushels per acre (Bu/A), compared with an average of about 53 Bu/A for the elite parental lines. Thus the high oleic transgenic soybeans are competitive in terms of yield with commercial elite soybean lines. In all field trials, the total oil content of the transgenic seeds was similar to those of normal seeds (Kinney, 1995).

It is most reassuring to know that it is possible to drastically alter the fatty acid content of a seed oil by genetic engineering without affecting the environmental stability or the seed of an elite plant.

10.5 Technological properties of high oleic acid soybeans

Unsaturated fatty acids are subject to autoxidation which causes oils containing them to become rancid. Autoxidation is initiated by free radicals or photoactivated oxygen and propagated by molecular oxygen (Frankel, 1980). The primary products of autoxidation are hydroperoxides, which can be quantified by the classic 'peroxide value' (PV) method. In this technique, hydroperoxides are reacted with iodide ions to form iodine which can then be titrated with thiosulfate. The resulting PV, in milliequivalents of iodine per kg of oil, is therefore a measure of the extent of fatty acid oxidation. For soybean oil, a peroxide value greater than 10 meq/kg has been correlated with unacceptable flavor.

After extended periods of time, or at high temperatures, hydroperoxides decompose into a variety of volatile and non-volatile secondary compounds including aldehydes and ketones (Frankel, 1980). These products are generally polar and may be quantified by high pressure liquid chromatography (HPLC). The same technique can be used to quantify the compounds formed by polymerization of volatile and non-volatile oxidation products formed at high temperatures (Lin, 1991). Volatile hydroperoxide breakdown products, such as hexanal and 3,6-nonadienal, are the cause of disagreeable flavors and odors of rancid oil and the food cooked in it. Thus, these volatiles can also be detected by human noses and taste buds and analysed qualitatively by sensory evaluation panels.

Of the many standardized methods used to evaluate the stability of commercial cooking oils, the most common is the Active Oxygen Method (AOM). This is an accelerated oxidation test in which an oil is aerated under a constant, elevated temperature (97.8°C) and degradation is monitored by measuring peroxide accumulation. The end point, or induction time, is determined by the number of hours required to reach a peroxide value of 100 meq/kg. Thus the longer the induction time the more stable the oil. Almost all commercial oil samples specify an AOM induction time as a component of the technical data sheet.

High oleic acid soybean oil has an AOM induction time of about 150 hours compared with 15 hours for normal soybean oil and about 200 hours for heavy duty shortening. Under standard AOM conditions, normal soybean oil reaches a PV of 10 meq/kg, and thus an unacceptable flavor, after about 7 hours. High oleic acid soybean oil does not reach this value until beyond 15 hours.

Incremental increases in either the linolenic or linoleic acid contents significantly reduce the AOM induction time of high oleic acid soybean oil. For example, when the linolenic acid content is increased to just 4%, the AOM induction time drops to about 100 hours. Thus it is extremely important that high oleic acid soybeans are completely separated from commodity soybeans at all stages of their production, from the field through crushing and processing. Further variations of the induction time are observed depending on the concentration of other minor components in the oil which affect stability.

Another standard method now commonly used to evaluate the stability of commercial cooking oils in the USA is the Oxidative Stability Index (OSI) measured automatically with a machine manufactured by Ominion, Inc, of Rockland, MA, USA. In the OSI machine, air is bubbled through oil heated to 110°C. As the oil oxidizes, volatile organic acids, primarily formic acid, are formed which can be collected in distilled water in a cell. The machine constantly measures the conductivity of the distilled water and the induction period is determined as the time it takes for this conductivity to begin a rapid rise. Although the data derived from the two methods do not always have a straight correlation, the OSI induction time values for most oils are generally about half those of the AOM derived values.

The OSI induction time of high oleic acid soybean oil is about 81 hours. This is over 10 times longer than normal soybean oil (7 hours) and close to the OSI induction time of solid hydrogenated shortening containing antioxidants (85 hours). High oleic acid soybean oil is also more stable in the OSI machine than commercial soybean oils containing the antioxidants TBHQ and silicone (21 hours) as well as liquid, partially hydrogenated soybean frying oil containing the same antioxidants (32 hours). As is true for the AOM values, the OSI induction times are significantly decreased by incremental increases in linoleic and linolenic acid contents above about 3%.

Somewhat surprisingly, the OSI induction time of high oleic acid soybean oil is remarkably higher than that of high oleic acid sunflower oil (18 hours). This is interesting since the fatty acid composition of these two oils is very similar and thus their theoretical oxidative potentials, determined by the method of Fatemi and Hammond (1980), are comparable (about 1.5).

Tocopherols are known to exert a strong antioxidant effect and these two high oleic acid oils do have different tocopherol contents and profiles. When tocopherol homologs are added to high oleic acid sunflower to match both the total content and the relative percentage of individual tocopherols in soybean oil, the OSI induction time is increased to 75 hours. When tocopherols are added such that only the total content is matched, and not the relative proportion of alpha, gamma and delta tocopherols, a smaller effect is observed (61 hours). Thus it appears that the stability of high oleic acid soybean oil is a result not only of the degree and type of fatty acid unsaturation but also of the relative proportion of individual tocopherols and of the total tocopherol content.

Although induction times obtained using the AOM and OSI techniques are widely used in product specifications, it is sometimes argued that they do not reflect a completely accurate estimate of the storage life of an oil or its behavior during frying. This is because the temperatures used by these methods are too high to simulate actual storage conditions (too many additional volatiles are formed not normally seen in stored oils) and too low to simulate frying conditions (many additional compounds are formed at the much higher frying temperatures).

One generally accepted method of measuring the storage life of an oil is the Schaal oven method. In this method, oil undergoes accelerated aging in a forced draft oven at 63°C. Oxidation can then be measured by PV, chromatographic analysis of the volatile compounds formed or by sensory evaluation of the oil. As might be expected, high oleic soybean oil develops peroxides at a far slower rate than normal soybean oil when maintained in a Schaal oven. After 200 hours the PV of the high oleic oil is still less than 5 meq/kg whereas normal soybean oil develops a PV of around 25 meq/kg after the same time period. Furthermore, even after 16 days in the Schaal oven, high oleic acid soybean oil still has an acceptable flavor, judged as somewhat grassy by a sensory panel. After the same period of time in the oven unmodified soybean oil tastes like paint and will have degraded to a level judged unacceptable by tasters.

To replicate frying temperatures, oils can be heated to around 190°C in glass tubes in temperature-controlled aluminum heating blocks. Usually, oils are heated for 10 hours per day and allowed to cool before samples are taken for HPLC analysis and the oils then reheated. When heated to 190°C over a period of several days, the accumulation of polar and polymer oxidation products in the high oleic acid soybean oil is far less than that of normal soybean oil (Figure 10.5). Indeed, the high oleic acid soybean oil has

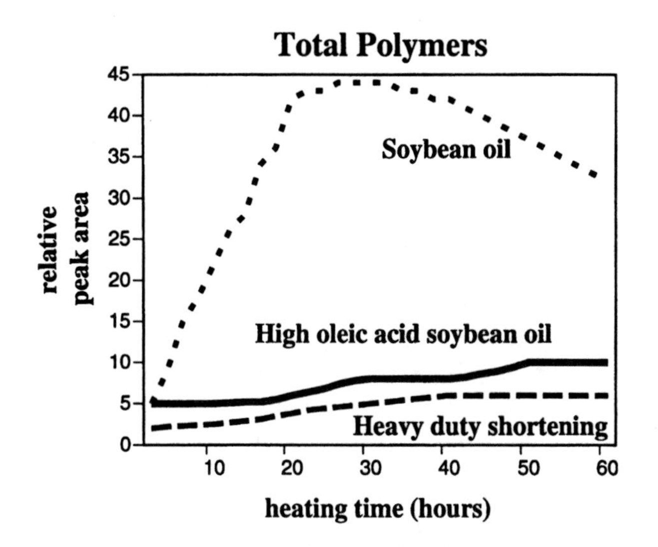

Fig. 10.5 Accumulation of polymerized oxidation products in soybean oils heated at 190°C for 0 to 60 hours. The rates of formation of polar products for the three oils are similar to the accumulation rates of the polymers shown here.

a similar performance at frying temperatures to that of heavy duty, solid shortening. It is worth noting that both of these oils developed far less polar and polymer material than high oleic sunflower oil, commercial clear liquid shortenings or soybean frying oils heated under the same conditions.

Sensory evaluation of food cooked in high oleic acid soybean oil is still in progress. Based on the above analysis, it is anticipated that the flavor quality of food cooked in high oleic acid soybean oil will be at least comparable to that of food cooked in heavy duty shortening.

When all these tests are taken in account, it is clear that the oxidative properties of high oleic acid soybean oil are considerably better than those of partially hydrogenated liquid oils and in some applications may be comparable to heavy duty solid shortenings (for further analysis see Knowlton *et al.*, 1997). However, unlike the chemically treated oils, the high oleic acid oil from GM soybean has a reduced content of palmitic acid and does not contain any *trans* fatty acids.

Epidemiological and metabolic studies during the past few decades have indicated that the consumption of different types of fatty acids can affect the ratios of low-density lipoproteins (LDL) to high density lipoproteins (HDL) in the blood plasma (Spady *et al.*, 1993). The concentration of cholesterol in the LDL fraction of plasma is considered to be one of the major risk factors for coronary heart disease (Wollett and Dietschy, 1994). Extensive studies have linked the consumption of saturated fats by animals and humans to coronary heart disease (McNamara, 1992; Allison *et al.*, 1995).

This is presumably since they increase the plasma concentration of total and LDL-cholesterol and/or reduce the concentration of beneficial HDL-cholesterol (McNamara, 1992; Spady *et al.*, 1993; Katan *et al.*, 1995). However, not all saturated fatty acids have a similar effect on plasma lipoproteins. For example, most studies report that stearic acid (18:0) has a neutral effect on the plasma concentration of LDL-cholesterol whereas palmitic acid (16:0), myristic acid (14:0) and lauric acid (12:0) appear to increase plasma LDL-cholesterol (Mensink *et al.*, 1994; Woollett and Dietschy, 1994). Thus, the reduced palmitic acid content of the high oleic acid soybean oil, compared with currently available hydrogenated and non-hydrogenated soybean oils, may have an improved nutritive value.

Consumption of *trans* fatty acids by humans has also been linked to unfavorable plasma lipoprotein profiles and coronary heart disease (see Kinney, 1996). There have also been reports of other adverse physiological effects of *trans* fatty acids when they are metabolized or incorporated into membranes (see Holman *et al.*, 1983). Unfortunately, the data for *trans* fatty acids are more equivocal than those for palmitic acid, possibly because of the difficulties involved in quantifying *trans* fatty acid consumption (Allison *et al.*, 1995) or because of the effects of interactions of *trans* fatty acids with dietary cholesterol or polyunsaturated fatty acids (Dictenberg *et al.*, 1995; Bolton-Smith *et al.*, 1996). It is apparent, however, that the unresolved issues surrounding the consumption of *trans* fatty acids are not a concern when using the high oleic acid soybean oil. Thus, it has been shown that it is now possible to create a soybean frying oil with the functionality of hydrogenated oils and the improved nutrition of a lower palmitic acid content while at the same time avoiding the potential health hazards from *trans* fatty acid consumption.

It is likely too that the meal from high oleic acid soybeans will have improved functionality in food applications because of the changes in the storage protein profile. The two major storage proteins in soybean are the 7S and 11S globulins which account for about 70% of total meal protein (Kinsella, 1979). The 7S fraction is made up of the a, a′ and b subunits of b-conglycinin. The 11S fraction is made up of the acidic (A) and basic subunits (B) of glycinin. The two globulins have considerably different effects on the functional aspects of the soy proteins in food (Kinsella, 1979; Kitamura, 1995) and it has been suggested that increasing the content of glycinin (11S) subunits and decreasing the content of b-conglycinin (7S) subunits can improve the functionality of soy proteins in various foods (Kitamura, 1995). In the high oleic acid soybeans the concentration of b-conglycinin a and a′ subunits are reduced and replaced with A and B subunits of glycinin. Preliminary functionality studies indicate that protein isolates from the meal of high oleic acid soybeans do indeed have better thermal stability, gel-making ability and emulsifying capacity in emulsified meat products than standard soybean protein isolates.

10.6 Health, safety and legislative issues

The Food and Agriculture Organization of the United Nations (FAO) and the World Health Organization (WHO) have defined food safety as 'providing assurance that food will cause no harm to the consumer when it is prepared and/or eaten according to its intended use' (FAO/WHO, 1995). A joint FAO/WHO consultation established that an important method of assessing safety of a new food is by comparing it with one already having an acceptable standard of safety (WHO, 1991). Thus, a new food may be considered safe if it is substantially equivalent to an existing food.

This concept of 'substantial equivalence' has also been advocated by the Organization for Economic Cooperation and Development (OECD) as the most practical means of evaluating new foods resulting from biotechnology (OECD, 1993). The type of safety assessment appropriate for a new food therefore depends upon the type of food being evaluated and the available knowledge of existing foods of a similar nature to the new food. A recent FAO/WHO consultation (FAO/WHO, 1996) discusses three possibilities for genetically engineered foods which result from this comparative approach: genetically engineered food that is substantially equivalent to existing foods, genetically engineered food that is substantially equivalent to existing foods except for defined differences and, finally, genetically engineered food that is not substantially equivalent to existing foods (for further discussion of these issues, see Chapter 4).

The composition of high oleic acid transgenic soybeans was compared with the elite parent variety in order to determine if the new beans could be considered substantially equivalent to regular soybeans. The transgenic soybeans were from two locations, Iowa and Puerto Rico, grown during the summer of 1995 and the winter of 1995/1996. The seeds represented the R4 and R5 generation of the high oleic acid transgenic soybeans lines described above. The composition analysis included determination of proximate, fatty acid, amino acid, isoflavone, raffinose, stachyose, phytic acid, and trypsin inhibitor content.

Proximate analysis included the measurement of protein, oil, crude fiber, carbohydrate and ash content. High oleic soybeans were indistinguishable from commodity soybeans in proximate composition.

Soybeans are valued as a major source of a high-protein animal feed supplement and have increasingly important utility in human food applications. We therefore wanted to establish that there was no change in the amino acid composition of high oleic lines. High oleic acid transgenic soybeans did not differ substantially from normal beans in any of the 17 amino acids measured.

As described previously, high oleic acid transgenic soybeans differ substantially from commodity soybean in the level of oleic, 9,12-linoleic, 9,15-linoleic, linolenic, and to a lesser extent, palmitic acid present in the oil.

However, with the exception of the 9,15 isomer, the relative abundance of minor fatty acids was similar in elite and transgenic beans.

The 9,15 isomer was present in the transgenic oil at less than 1% of the total fatty acid content. This isomer is also found, at concentrations ranging from 0.02% to 5.4% of the total fatty acids, in many edible sources of fat including butterfat, cheese, beef and mutton tallow, partially hydrogenated vegetable oils, human milk and mango pulp (De Jong and van der Wel, 1964; Keppler *et al.*, 1965; Stroink and Sparreboom, 1967; Hoffman and Meijboom, 1969; Murawski *et al.*, 1971; Mallet *et al.*, 1985; Ratnayake and Pelletier, 1992; Werner *et al.*, 1992; Shibahara *et al.*, 1993). In our own studies, we observed that commercially available hydrogenated soybean oil contained about twice the mol % of 9,15-isomer than high oleic oil from transgenic soybeans.

Several heat labile and heat stabile 'antinutritional' factors are known to exist in soybean including trypsin inhibitors, phytic acid, and the oligosaccharides raffinose and stachyose. Trypsin inhibitors are heat labile and are destroyed during the processing of soy protein products by heat treatment. They are associated with hypertrophy and lesions of the pancreas and the loss of S-containing amino acids which limits animal growth. Phytic acid is heat stable and has been implicated in interfering with the bioavailability of minerals such as calcium, magnesium and zinc. Phytic acid remains stable through most soybean processing steps. Raffinose and stachyose are associated with flatulence resulting from ingestion of soybean flours. Further processing of soybean flours into concentrates and isolates removes these oligosaccharides. High oleic acid transgenic and elite soybeans were analysed for each of these constituents and no differences were observed in any of these components.

Soybeans contain several biologically active phytoestrogens known as isoflavones which are present in various concentrations in soy protein products. Isoflavone content varies between soybean varieties and is affected by environmental factors. Minimally processed soy products contain higher levels of isoflavones than do products that undergo an alcohol wash as part of their purification. Recent studies suggest that the isoflavones may act as cancer-protective agents since they have been linked to a variety of anticarcinogenic activities including antioxidant, radical scavenging, serum cholesterol lowering and antiestrogenic and antiproliferative properties (Wang and Murphy, 1994; Messina *et al.*, 1994). The major isoflavones in soybeans and soybean products include daidzin, genistin, and their corresponding aglycons, daidzein, genistein. Glycitin and glycitein also occur in trace amounts (Wang *et al.*, 1990). The concentration of total genistein in both the high oleic acid transgenic soybeans and the elite controls were similar and a little below the reported ranges. No differences were observed between control and high oleic soybean in either total daidzein or glycitein content.

As described previously, there was a reduction in the concentration of the a and a′ subunits of b-conglycinin in the high oleic acid transgenic soybeans when compared with elite beans. This was coincident with an increase in the concentration of the acidic and basic subunits of glycinin in the transgenic soybeans, in addition to an increase in the concentration of the A2B1A glycinin precursor.

Based on the analytical evidence described above, high oleic acid transgenic soybeans at first appear to fall into the second FAO/WHO category. That is, they are substantially equivalent to existing soybeans except for clearly defined differences in their oil fatty acid composition and in the ratio of storage proteins in their meal fraction.

Once a new food has been determined to be substantially equivalent, the FAO/WHO consultation concluded that further safety assessment should focus only on these defined differences (FAO/WHO, 1996). In the case of high oleic acid transgenic soybeans the defined differences in fatty acid composition are themselves substantially equivalent to other, existing vegetable oils that are generally recognized as safe (e.g. high oleic acid sunflower oil). The linoleic acid isomer is absent from non-hydrogenated soybean oil and high oleic sunflower oil but is present, at ranges from 0.02% to 5.4% of the total fatty acids, in butterfat, beef and mutton tallow, partially hydrogenated vegetable oils, human milk and mango pulp (for references, see previous section). Thus, the amount observed in high oleic acid transgenic soybeans (<1%) is well within the range observed and considered safe in existing foods.

Although the storage protein ratio was different from the parent soybean line, it was similar to profiles found in other existing soybean varieties used for both animal feed and human foods. Therefore, no allergenicity problems were anticipated. Nevertheless, allergenicity testing was carried out by Dr Samuel Lehrer at Tulane University School of Medicine in New Orleans. Sera from 31 subjects with a documented history of soybean or food allergy, plus a positive skin test to soybean extract, and/or a positive IgE antibody response to soybean extract, were used. Control sera were obtained from soybean tolerant individuals. The subjects gave informed consent and completed a questionnaire concerning their allergy history, reactivity to soy and skin test reactivity to common inhalant and food allergens. All sera were assayed for soybean-specific IgE antibody by the established radioallergosorbent test (RAST) using discs coated with increasing amounts of either elite or transgenic soybean extracts.

The results indicated that both extracts yielded identical RAST inhibition. Both inhibition curves were analysed by logit-log transformation and linear regression analysis. The two regression lines were compared for statistically significant differences of their slopes and y-axis intercepts. The correlation coefficients for both elite and transgenic soybean were high ($r_{wt} = 0.9991$, $r_t = 0.9978$) and the 50% inhibition concentrations were very similar ($I_{50,wt}$ = 16.7 μg/ml, $I_{50,t}$ = 11.7 μg/ml). Analysis showed that both slopes and y-axis intercepts were statistically identical. Thus, from a quantitative viewpoint, the allergen content of both soybean extracts was identical. Twenty-one of the most potent RAST positive sera were selected for immunoblot (Western) analyses of soybean allergens. The results demonstrated that there were no significant differences in the number of bands to which the sera reacted or the intensity of the IgE reactivity between the transgenic soybean extract and the elite soybean extract. Thus, it was demonstrated that there was no significant quantitative or qualitative difference between the transgenic and elite soybeans with regard to their allergen content.

Based on the guidelines of the WHO, FAO and the OECD, oil and meal from high oleic acid transgenic soybeans can be considered substantially equivalent to existing foods and are therefore safe.

In the United States, the primary agency responsible for ensuring the safety of bioengineered foods from plants is the Food and Drug Administration (FDA). The policy of the FDA regarding transgenic plants is outlined in a recent statement (FDA, 1992) and is also covered in Chapter 4. In general, anyone wishing to commercialize a transgenic crop plant in the USA is required to consult with FDA. The high oleic acid transgenic soybeans were evaluated according to the guidelines and flow charts in the FDA statement (FDA, 1992) in full consultation with the FDA throughout the development of the product. In 1996, the FDA had no further questions regarding the safety of high oleic acid transgenic soybeans and confirmed this in writing in April 1997. Thus, this crop has now been added to the FDA's list of bioengineered foods derived from new plant varieties and may be developed commercially in the USA.

10.7 Environmental issues

Two of the most commonly expressed environmental concerns regarding transgenic crop plants are the movement of transgenes into wild populations by outcrossing and the horizontal transfer of antibiotic resistance genes into microbial populations. The former concern is irrelevant in the case of transgenic soybeans grown in the USA and Europe since there simply are no wild relatives of the soybean with which the crops could outcross. Furthermore, even when they are in close proximity to each other, soybeans do not outcross anyway (the rate of cross-fertilization in a field of soybeans is less than 1%). Moving pollen from one soybean to another is an art mainly restricted to plant breeders.

The high oleic acid transgenic soybeans contain a gene for bacterial ampicillin resistance (the *bla* gene, encoding a beta-lactamase enzyme) which is not expressed in the plant. A joint FAO/WHO expert consultation on biotechnology and food safety has recently concluded that the possibility of transfer of such a gene from a transgenic plant to a microorganism in the environment is vanishingly small (FAO/WHO, 1996), a conclusion which is in complete concurrence with the scientific evidence (Prins and Zadoks, 1994; Schlueter *et al.*, 1995). Furthermore, even if this 1-in-500 000 000 000 000 000 event did occur, the number of bacteria acquiring the gene would be infinitesimally small compared with the number of microorganisms in the world already containing the gene. Finally, it should be noted that the *bla* gene confers resistance only to a primitive form of ampicillin no longer used in clinical practice (Salyers, 1995). The *bla* gene is not effective in conferring resistance to the modern generation of penicillins and cephalosporins nor is it resistant to beta-lactamase inhibitors such as sulbactam and clavulanic acid (Salyers, 1995). Thus it is clear that the presence of the *bla* gene in a transgenic plant cannot rationally be considered a safety or environmental issue.

Since the high oleic acid transgenic soybeans also contain some DNA from *Agrobacterium tumefaciens* and cauliflower mosaic virus, they are considered regulated articles by the USDA. Any time that the seeds from these plants are transported or grown within the USA, the Animal and Plant Health Inspection Service (APHIS) of the USDA must be notified. Based on the information described above, and on extensive field trials of the new soybean lines, we expect that the high oleic acid transgenic soybeans will achieve a nonregulated status in 1997. This means that they will be exempted from APHIS notification and will be treated in the same manner as non-transgenic soybean varieties.

10.8 Labelling of the high oleic acid soybean

The labelling and legislative requirements for marketing food from genetically engineered plants vary from country to country and may vary from product to product. The new European Union (EU) Regulation on novel foods, for example, will require labelling if the new food is substantially different from conventional foods (Johnson, 1997; see also Chapter 4). It might be assumed, then, that oil from a transgenic soybean would only require labelling if it differed from conventional soybean oil, since the oil itself does not contain any transgenes or novel proteins. Since the new soybean oil described here is quite unlike any existing soybean oil, a label such as 'high oleic acid soybean oil' might be appropriate within the EU. However, approval for such a label would be dealt with on a case-by-case basis by EU Scientific and Standing Committees. It is possible that such committees might also require that the label outlines the process by which the oil was produced (see Johnson, 1997, for discussion).

In the USA, the FDA considers genetic engineering to be an extension of traditional methods of plant variety development (FDA, 1992) and does not consider the method by which a new plant variety is developed relevant to the labelling of the food product derived from that variety. This is logical when it is considered that modern plant varieties are created by using one or more of a number of different techniques including chemical mutagenesis and tissue culture. In these cases, the consumer is generally unconcerned about how those varieties were made. As a general principle, however, DuPont supports the consumers' right to know about the food they eat. When it is important to consumers, or in those areas where it is required by law, DuPont will inform customers of the plant breeding technique used to produce a specific value-enhanced product. United States law does require that foods derived from new varieties which are substantially different from their traditional counterparts be labelled in a manner informative to the consumer. For these reasons it is likely that oil from the transgenic soybeans described above will be labelled in the US as 'high oleic acid soybean oil' to distinguish it from commodity soybean oil. Since the high oleic acid soybeans are 'identity preserved', it will be possible for consumers to distinguish products derived from transgenic beans from those of non-transgenic bean products, if they so desire.

10.9 Consumer acceptance and marketing

Since the market for edible soybean oil in the USA is more than 10 billion pounds, it is apparent that the replacement of hydrogenated commodity oils by high oleic acid soybean will have to be gradual. Initial markets will be specialized, providing replacements for current speciality oils such as high oleic acid sunflower oil. As the acreage increases, it will be possible to supply broader markets, such as frying oils for french fries or potato chips. Finally, it is possible that high oleic acid soybean will have the potential to replace the current soybean oil as the standard, commodity oil. Given the perceived health benefits of the new oil and of foods cooked in it, it is anticipated that consumers will welcome a widely available alternative to chemically hydrogenated vegetable oils.

10.10 Future prospects: true designer oils

Both high oleic acid soybean and high lauric acid rapeseed oils are currently being commercialized. High stearic acid oils are not far behind. A number of other new oils, such as zero saturate soybean oil (Kinney, 1996) and oils containing γ-linolenic acid (Reddy *et al.*, 1996) are currently being developed. What then does the future hold beyond these products? Two overlapping areas of research that will have a significant impact in the next century on the types of genetically engineered edible oils that will be available are 'designer enzymes' and 'designer molecules'.

As more is understood about the nature of fatty acid biosynthetic enzymes, it is becoming possible to modify them in a way that will change the types of fatty acids found in the seed oils. One way to make unusual fatty acids in domestic crops is to isolate the appropriate gene from an exotic plant which makes the unusual fatty acid and express this heterologous gene in the desired crop. It is now becoming possible to take existing genes from domestic crop plants and modify them so that they make enzymes similar to ones found in exotic plants. It will even be possible to create enzymes which make fatty acids that are not found in known plants. For example, Calgene have begun to modify the specificity of enzymes called thioesterases by protein engineering (Yuan *et al.*, 1996). As a first step, they have modified specificity of their California Bay Laurel thioesterase so that it produces 14-carbon fatty acids rather than 12-carbon fatty acids. Expression of this new gene in an oilseed results in oil rich in 14-carbon fatty acids (Yuan, 1996).

Even more impressively, it is now becoming possible to modify desaturase enzymes so that they will insert double bonds in carbon chains of different lengths at a specified position in a carbon chain (Shanklin *et al.*, 1997). Oils containing these fatty acids can then have the exact melting point or SFC desired for a particular application.

It may soon be possible to engineer the enzymes which attach fatty acids to the glycerol backbone (acyltransferases) so that they will put a desired fatty acid only at a specified position on the molecule. Not only will it then be possible to have an oilseed plant make exactly the classes of fatty acids required but also have it put these fatty acids at specific positions on the triacylglycerol molecule. This new designer molecule, or structured triacylglycerol, will then have exactly the right functional properties required for a desired food application.

Existing structured triacylglycerols are currently made only by expensive transesterification and fractionation methods. Genetically engineered structured triacylglycerols will be the true 'designer oils'. They will have a wide range of applications from cocoa butter substitutes with exactly the required mouthfeel, to infant formulas, intravenous feeds, athletic supplements and perhaps dozens of other applications not yet conceived.

References

Allison, D. B. *et al.* (1995) Trans fatty acids and coronary heart disease risk. *Am. J. Clin. Nutr.*, 62: 655–707.

Ascherio, A. and Willett, W. C. (1995) New directions in dietary studies of coronary heart disease. *J. Nutr.*, 125: 647–55.

Bolton-Smith, C. *et al.* (1996) Does dietary fatty acid intake relate to the prevalence of coronary heart disease in Scotland? *Euro. Heart J.*, 17: 837–45.
Craig-Schmidt, M. C. (1992) Fatty acid isomers in foods, in *Fatty Acids in Foods and Their Health Implications* (C. K. Chow, ed.) Marcel Dekker, Inc., New York, pp. 365–98.
De Jong, K. and van der Wel, H. (1964) Identification of some iso-linoleic acids occurring in butterfat. *Nature*, 202: 553–5.
Dictenberg, J. B., Proczuk, A. and Hayes, K. C. (1995) Hyperlipidemic effects of trans fatty acids are accentuated by dietary cholesterol in gerbils. *J. Nutr. Bichem.*, 6: 353–61.
Fatemi, S. H. and Hammond, E. G. (1980) Analysis of oleate, linoleate and linolenate hydroperoxides in oxidised ester mixtures. *Lipids*, 15: 379–85.
Del Vecchio, A. J. (1996) High laurate canola. *Inform*, 7: 230–43.
FAO/WHO (1995) Report of the 28th session of the Codex Committee on food hygiene, FAO, Rome.
FAO/WHO (1996) Biotechnology and food safety. Report of a Joint FAO/WHO Consultation, FAO, Rome.
FDA (1992) Statement of policy: Foods derived from new plant varieties. *Federal Register* 57(104): 22984–3004.
Frankel, E. N. (1980) Lipid oxidation. *Prog. Lipid Res.*, 19, 1–22.
Heppard, E. P. *et al.* (1996) Developmental and growth temperature regulation of two different microsomal omega-6 desaturase genes in soybean. *Plant Physiol.*, 110, 311–19.
Hoffman, G. and Meijboom, P. W. (1969) Identification of 11,15-octadecadienoic acid from beef and mutton tallow. *JAOCS* 46: 620–2.
Holman, R. T. *et al.* (1983) Metabolic effects of isomeric octadecanoic acids, in *Dietary Fats and Health* (E. G. Perkins and W. T. Visek, eds) Am. Oil Chem. Soc., Champaign, Illinois, pp. 320–40.
Johnson, E. (1997) European politicos pass food regulations lacking meat. *Nature Biotech.*, 15: 210.
Jorgensen, R. A. and Napoli, C. A. (1994) Genetic engineering of novel plant phenotypes. US Patent 503 4323. (Priority date March 1989.)
Katan, M. B., Zock, P. L. and Mensink, R. P. (1995) Dietary oils, serum lipoproteins and coronary heart disease. *Am. J. Clin. Nutr.* 6: 1368–73.
Keppler, J. G. *et al.* (1965) Components of the hardening flavor present in hardened linseed oil and soybean oil. *JAOCS*, 42: 246–9.
Kinney, A. J. (1994) Genetic modification of the storage lipids of plants. *Curr. Opin. Biotechnol.* 5: 144–51.
Kinney, A. J. (1995) Improving soybean seed quality, in *Induced Mutations in Molecular Techniques for Crop Improvement.* International Atomic Energy Agency, Vienna, Austria, pp. 101–13.
Kinney, A. J. (1996) Development of genetically engineered soybean oils for food applications. *J. Food Lipids*, 3: 273–92.
Kinsella, J. E. (1979) Functional properties of soy proteins. *JAOCS*, 56: 242–58.
Kitamura, K. (1995) Genetic improvement of nutritional and food processing quality in soybean. *JARO*, 29: 1–8.
Knowlton, S., Ellis, S. K. B. and Kelley, E. F. (1997) Oxidative stability of high oleic acid soybean oil. *JAOCS*, in press.
Knutzon, D. S. *et al.* (1992) Modification of brassica seed oil by antisense expression of a stearoyl-acyl carrier protein desaturase gene. *Proc. Natl. Acad. Sci.* (USA), 89: 2624–8.
Lin, S. S. (1991) Fats and oils oxidation, in *Introduction to Fats and Oils Technology* (P. J. Wan, ed.) Am. Oil Chem. Soc., Champaign, Illinois, pp. 211–32.
Mallet, G. *et al.* (1985) Studies on the mechanism of selective hydrogenation. Structures of isomers formed by hydrogenation of a model simulating rapeseed oil. *Rev. Fr. Corps Gras* 32: 387–95.
Mensink, R. P., Temme, E. H. M. and Hornstra, G. (1994) Dietary saturated and trans fatty acids and lipoprotein metabolism. *Annal. Med.*, 26: 461–4.
Messina, M. J., *et al.* (1994) Soy intake and cancer risk – a review of the in-vitro and in-vivo data. *Nutrition And Cancer – An International Journal*, 21: 113–31.
McNamara, D. J. (1992) Dietary fatty acids lipoproteins and cardiovascular disease, in *Advances in Food and Nutrition Research* Vol. 36 (J. E. Kinsella, ed.) Academic Press, Inc., San Diego, pp. 254–351.

Murawski, U. *et al.* (1971) Identification of non-methylene-interrupted cis,cis-octadecadienoic acids in human milk. *FEBS Lett.*, 18: 290–2.
OECD (1993) *Safety evaluation of foods produced by modern biotechnology: Concepts and principles*. OECD, Paris.
Ohlrogge, J. B. (1994) Design of new plant products: Engineering of fatty acid metabolism. *Plant Physiol.*, 104: 821–6.
Prins, T. W. and Zadoks, J. C. (1994) Horizontal gene-transfer in plants, a biohazard – outcome of a literature-review. *Euphytica*, 76: 133–8.
Ratnayake, W. M. N. and Pelletier, G. (1992) Positional and geometrical isomers of linoleic acid in partially hydrogenated oils. *JAOCS*, 69: 95–105.
Reddy, A. S. and Thomas, T. L. (1996) Expression of a cyanobacterial D6-desaturase gene results in γ-linolenic acid production in transgenic plants. *Nature Biotechnology*, 14: 639–42.
Salyers, A. A. (1995) *Gene Transfer in the Mammalian Intestinal Tract: Impact on Human Health, Food Safety and Biotechnology*. R. G. Landes Co. Biomedical Publishers, Austin, TX.
Schlueter, K., Fuetterer, J. and Potrykus, I. (1995) Horizontal gene transfer from a transgenic potato line to a bacterial pathogen occurs, if at all, at an extremely low frequency. *Biotechnology*, 13: 1094–8.
Shanklin, J. *et al.* (1997) Structure-function studies on desaturase and related hydrocarbon hydroxylases, in *Physiology, Biochemistry and Molecular Biology of Plant Lipids* (J. P. Williams, M. U. Khan and N. W. Lem, eds) Kluwer Academic Publishers, Dordrecht, pp. 6–10.
Shibahara, A. *et al.* (1993) cis-9,cis-15-octadecadienoic acid: a novel fatty acid found in higher plants. *Biochim. Biophys. Acta*, 1170: 245–52.
Spady, D. K., Woollett, L. A. and Dietschy, J. M. (1993) Regulation of plasma LDL-cholesterol levels by dietary cholesterol and fatty acids. *Ann. Rev. Nutr.*, 13: 355–81.
Stroink, J. B. A. and Sparreboom, S. (1967) Synthesis of cis 9,15-, 8,15- and 7,15-octadecadienoic acid. *JAOCS*, 44: 531–3.
Wang, G. *et al.* (1990) A simplified method for the determination of phytoestrogens in soybean and its processed products. *J. Agric. Food Chem.*, 38: 185–90.
Wang, H. and Murphy, P. A. (1994) Isoflavone content in commercial soybean foods. *J. Agric. Food Chem.*, 42: 1666–73.
Werner, S. A., Luedecke, L. O. and Shultz, T. D. (1992) Determination of conjugated linoleic acid content and isomer distribution in three cheddar-type cheeses effects of cheese cultures processing and aging. *J. Agric. Food Chem.*, 40: 1817–21.
WHO (1991) Strategies for assessing the safety of foods produced by biotechnology. Report of a Joint FAO/WHO Consultation, WHO, Geneva.
Wiseman, M. J. (1992) Present and past trends in dietary fat consumption, in *Dietary Fats: Determinants of Preference, Selection and Consumption* (D. J. Mela, ed.), Elsevier, London, pp. 1–8.
Wollett, L. A. and Dietschy, J. M. (1994) Effect of long-chain fatty acids on low-density-lipoprotein-cholesterol metabolism. *Amer. J. Clin. Nutr.*, 60: 991–6.
Yadav, N. S. *et al.* (1993) Cloning of higher plant omega-3 fatty acid desaturases. *Plant Physiol.* 103, 467–76.
Yuan, L. (1996) Engineering plant thioesterases and disclosure of plant thioesterases having novel substrate specificity. International Patent Application WO9636719.
Yuan, L., Voelker, T. A. and Hawkins, D. J. (1995) Modification of the substrate-specificity of an acyl-acyl carrier protein thioesterase by protein engineering, *Proc. Natl Acad. Sci.* (USA), 92: 10639–43.

11 Potatoes

DAVID M. STARK

11.1 Introduction and historical perspective

Potato is the number one vegetable and fourth largest crop in the world, with an annual farmgate value of over $35 billion. In the United States, potato generates $27 billion at retail with about 50% of this coming from French fries. From a scientific standpoint, potato has long served as a model system and learning tool to the biotechnology research community, primarily due to the ease with which transgenes can be inserted and transgenic plants obtained via tissue culture. While traditional potato breeding is very difficult due to high genetic heterogeneity, nearly all potato cultivars including all major commercial varieties are readily amenable to transformation and regeneration. In fact, potato is one of the fastest and most efficient systems for adding and testing transgenes. Given the ease of manipulation and importance as crop, it is not surprising that a wealth of information has been obtained from laboratories around the world working with transgenic potatoes.

Foreign genes are typically inserted into potato using *Agrobacterium*-mediated transformation techniques (for review, see Klee, 1987). The first transgenic potato was reported in 1986, and in 1995 the first transgenic potato was fully approved for sale in the United States, having undergone approval by the FDA, USDA and EPA. This potato is an improvement of Russet Burbank, the dominant fresh market and processing variety in North America, and contains a gene which confers complete resistance to the Colorado Potato beetle, the most devastating insect pest on potato in North America and much of the world. Trademarked NewLeaf™, the potato is sold through the NatureMark™ division of Monsanto and will be discussed further in this chapter. NatureMark™ has extended the insect control trait to other major varieties and is in various stages of commercial launch, and in 1998 will add resistance to potato leaf roll virus and potato virus Y to the product portfolio. Other future products will include resistance to major bacterial and fungal pathogens, and improvements in processing traits including high solids, improved storability and reduced blackspot bruise. This chapter will highlight results and initial reaction to the commercial introduction of the NewLeaf™ potato, and summarize progress to date in the area of processing trait improvements.

Several major companies in addition to Monsanto are actively pursuing biotechnology as a means of improving potato. These include Frito-Lay, which has invested in improving solids, storability and pest resistance. Danisco (Copenhagen) has developed a variety of technologies also aimed at improving quality traits such as high starch and improving storability by preventing starch breakdown at low temperatures. Advanced Technologies (Cambridge), Zeneca, Avebe and Calgene also have programs in similar areas of research.

11.2 Production process and patent status

11.2.1 NewLeaf™ potato

For over 40 years certain classes of insects have been controlled by a naturally occurring protein found in the common soil bacterium *Bacillus thuringiensis* (Bt) (for review, see Koziel, 1993). During the sporulation phase, the bacteria accumulate large quantities of crystalline proteins known as δ-endotoxins. These proteins are solubilized and processed in the insect gut into the active form of the toxin. The advantage of these toxins is that they are both highly active and highly specific to certain types of insects. Thousands of different strains of *Bacillus* have been identified resulting in a large array of distinct classes of δ-endotoxins. Since the toxins are proteins, the genetic information or gene for these proteins can easily be isolated and inserted into plants. The NewLeaf™ potato was made by insertion of the *Cry*IIIA gene, which encodes the δ-endotoxin found in *B. thuringiensis* subsp. *tenebrionis* (Btt) (Perlak, 1993). This protein is highly specific to certain coleopteran insects, including the Colorado potato beetle. The protein is not toxic to other types of insects, earthworms, birds, fish or mammals.

The Btt gene was first inserted into the potato variety Russet Burbank via *Agrobacterium*-mediated transformation. In this process of gene insertion, a marker must be used for selection of cells which contain the new gene. In this case, the marker was neomycin phosphotransferase (NPT II), which confers resistance to the antibiotic kanamycin. This marker has been approved for use in a number of different plant products in a number of countries, and this will be discussed below. Both the Btt and NPT II genes were expressed from the 35S promoter resulting in production of the transgenic proteins throughout the plant. At first, Btt accumulation was too low to provide effective insect control. Expression was improved to provide commercial levels of control by modifying the coding region to make the gene look, to the plant cell, like other plant genes (Perlak, 1993). The Btt protein accumulated to approximately 0.12% of the total leaf protein and less than 0.01% of the total tuber protein (Lavrik, 1995). The low level of protein in the tuber is expected given that the promoter is less active in driving gene expression in the tuber. The NPT II protein accumulated at 2–5-fold lower levels that that of the Btt protein. Again, the lower accumulation of the NPT II protein is expected given that the Btt gene was optimized for expression in plants.

11.2.2 High solids potato

Potatoes processed into fries and chips must meet certain quality specifications, including size, shape, solids content, sugar content and freedom from internal defects such as hollow heart and bruise. The trait with the greatest influence on product throughput and finished product quality is the solids content. A higher solids potato has value due to lower water content which contributes to reduced transportation costs, cooking time and costs, and increased processed-product recovery. In addition, oil absorbed during cooking is decreased resulting in a product with reduced calories. Starch is the major contributor to solids in tubers and thus enhancement of starch biosynthesis was targeted as a means of increasing dry matter content.

Starch biosynthesis in plants and glycogen biosynthesis in bacteria are essentially identical processes at the biochemical level (Preiss, 1991). In both instances, the rate-limiting step is catalysed by the enzyme ADPglucose pyrophosphorylase (ADPGPP). The activity of this enzyme is highly regulated by the metabolic state of the cell, and this regulation thus controls the amount of carbohydrate being stored in the form of starch. A strain of *Escherichia coli* was identified as being an overproducer of glycogen (Cattaneo, 1969). Further studies showed that the overaccumulation of glycogen was due to the presence of an ADPGPP enzyme which was no longer sensitive to metabolic regulation by the cell due to a single amino acid change in the protein sequence (Lee, 1987). The corresponding gene encoding this enzyme, called *glgC16*, was isolated and inserted into Russet Burbank potato plants via *Agrobacterium*-mediated transformation. Starch content of tubers expressing this ADPGPP variant is increased significantly, with a concomitant increase in dry matter content. On average, these tubers contain 24% higher levels of dry matter relative to controls. These results represented the first significant increase in carbohydrate content produced via biotechnology (Stark, 1992).

International and US patents have been filed and claims have been issued describing the use of ADPGPP enzymes from any source which is less sensitive to allosteric control for the purpose of increasing the starch content in plants. Related patents have also been issued whereby this same technology has proven useful for increasing the sugar content and overall flavor in tomatoes, decreasing oil in seeds, and improving the cold storage properties in potatoes. All of the patents are owned by the Monsanto Company.

11.3 Field/process performance

11.3.1 NewLeaf™ potato

NewLeaf™ potatoes have been grown experimentally since 1991 and commercially since 1995. In all, NewLeaf™ will be grown on nearly 35 000 acres in 1997. In every instance NewLeaf™ has provided season-long, 100% control of the Colorado potato beetle without the use of any insecticide (Table 11.1). In most other respects, NewLeaf™ performs exactly as any other Russet Burbank potato as determined through several growing seasons across North America. However, several other important characteristics of the transgenic potato have also probably been improved by clonal selection. Since potato varieties are propagated through cuttings, commercial varieties are constantly improved through clonal selection, or simply choosing those individual plants with the best performance. The plant giving rise to NewLeaf™ Russet Burbank actually shows a higher percentage of larger tubers which translates into a higher payable yield to the grower. Also, the tubers are more dormant and maintain overall quality better during prolonged storage.

Due to the lack of systemic insecticide applications for beetle control, beneficial insects such as spiders survive in the field and provide control of other insect pests, the most notable being the green peach aphid. Green peach aphids do not generally harm the plant *per se*, but rather cause damage through effective transmission of several viruses, including potato leaf roll virus and potato virus Y. These viruses can impact yield, quality and thus marketability of the potatoes, as well as causing severe losses in seed production areas due to de-certification for use as seed. In general, North American growers of NewLeaf™ potatoes were able to reduce total insecticide use by 42% due to complete control of Colorado potato beetle and partial control of green peach aphids due to higher populations of predatory insects (Table 11.2).

NewLeaf™ potatoes have been grown in Turkey and the Republic of Georgia. In both cases where beetle pressure was very high, NewLeaf™ performed as expected and provided complete beetle control. Since

Table 11.1 NewLeaf™ provides season-long, 100% control of Colorado potato beetles without insecticide applications. Data from field trials in Wisconsin

	% defoliation by Colorado potato beetle								
	Day 0	Day 7	Day 14	Day 21	Day 28	Day 35	Day 42	Day 49	Day 56
NewLeaf™ – no insecticide	0	0	0	0	0	0	0	0	0
Russet Burbank – insecticide	0	12	25	10	10	3	10	5	10
Russet Burbank – no insecticide	5	15	35	45	45	60	75	70	60

Table 11.2 Seasonal distribution of green peach and potato aphids. Hermiston, Oregon, 1992

	Total aphids per plot				
	Day 0	Day 17	Day 35	Day 52	Day 69
NewLeaf™ – no insecticide	0	20	30	30	0
Russet Burbank – systemic insecticide	0	30	5	5	0
Russet Burbank – foliar insecticide	0	170	1922	3969	1747

Colorado potato beetle is the number one insect pest in much of Eastern Europe and the former Soviet Union, plans are under way to introduce NewLeaf™ in a number of different potato varieties in this part of the world.

11.3.2 High solids potato

Transgenic high solids potatoes have been tested under field conditions since 1991 and in locations ranging across North America and recently in Scotland. In general, expression of the bacterial ADPGPP gene enhances starch biosynthesis and results in a 20–30% increase in the total solids content of the tuber with no reduction in fresh weight or in harvested yield (Table 11.3). Plant growth characteristics, plant phenotype, tuber size and tuber shape are normal as well. This technology thus represents a true improvement in crop productivity or harvested dry matter per acre. Fried product made from these potatoes was found as predicted to absorb proportionally less oil, have a crisper texture and a fuller potato taste.

Most potatoes destined for processing are stored, sometimes for as long as 10 months, in order to meet year-round demand at the processing factories. This storage poses a challenge for the grower and the processor as they must strike a balance between the desire to store at as low a temperature as possible to reduce shrinkage due to respiration, minimize sprouting and storage diseases, while maintaining fry quality by minimizing sugars which accumulate at these low temperatures. High sugar levels cause the

Table 11.3 Solids and yield in transgenic high solids potatoes. Data are from replicated plot field trials in Idaho. Cwt is hundredweight, or one hundred pound units

	% solids	Yield cwt/acre
Russet Burbank	18.9	374
Russet Burbank	19.2	351
High Solids – 1	21.3	413
High Solids – 2	21.9	381
High Solids – 3	22.3	304
High Solids – 4	23.1	347

potato to fry to an unacceptably dark color. Transgenic high solids tubers could withstand storage at 4°C for 4 months and produce product with acceptable color while controls fried quite dark (section 11.4). Lastly, the high solids tubers were more dormant than controls which is an advantage in that they would require fewer applications of sprout inhibitors during prolonged storage.

While the gain in solids and the dramatic improvement in storability are of significant scientific achievement and potential commercial value, the transgenic tubers are extremely susceptible to blackspot bruise. This type of bruise is very common in potato and is caused by mechanical damage which occurs during harvest and handling. Even though the potato industry is accustomed to dealing with minimizing blackspot bruise, the transgenic potatoes are too susceptible for normal handling practice and this negative trait is what has kept the high solids technology from further commercialization. However, technology for reducing blackspot bruise has been developed and this may allow for the high solids technology to actually become commercially available in the not-too-distant future

11.4 Technological properties

11.4.1 NewLeaf™ potato

Modification of the potato plant to include the constant synthesis of the insecticidal *B.t.t.* protein offers many advantages over current cultural practice. With traditional chemical controls, growers must constantly scout the field for insects, choose the correct insecticide and the correct timing (and hope the weather cooperates), spray the field, maintain the equipment and file the appropriate records. All of this continues throughout the growing season consuming a lot of time, effort and cost to the grower for what amounts to imperfect insect control, not to mention the insecticide load on the environment and in the grower's own backyard. Foliar applications of formulated *B.t.t.* products, while environmentally safer than many insecticides, have their own disadvantages. For example, sunlight will inactivate the protein, water washes it off the leaves and new growth is not protected by previous applications. In contrast, putting *B.t.t.* inside the plant through genetic modification makes sunlight, water and growth once again good for the plant. With NewLeaf™, the grower does not need to scout for beetles or otherwise worry about beetle control for the entire growing season.

The average grower saved $5/acre in total insecticide costs in 1996 by using NewLeaf™ versus standard Russet Burbank, with the absolute savings determined by geography and natural beetle pressure. These savings represent the difference between the technology license fee charged by NatureMark™ for the use of NewLeaf™ and the average cost of insect control in unimproved potato production. The average grower was able to reduce total insecticide use by 42% in 1996. In regions where beetle pressure was higher and normal chemical applications more frequent, the grower savings tended to be much higher. Of interest and significance to the industry, the payable yield was higher with NewLeaf™ than with standard Russet Burbank, accounting for an additional average benefit of $35/acre for the grower. The payable yield is related to the number of US grade one potatoes produced by the crop, and growers using NewLeaf™ were able to produce a much higher quality crop. In the end, the grower has applied fewer chemicals to the farm, spent less time killing beetles and more time managing the crop, saved money and produced a better quality and more valuable potato crop. An additional benefit beginning to be realized by growers is that NewLeaf™ maintains quality much better during storage, and this increased dormancy will result in better payout to the grower. The actual financial benefit to the grower is still being determined.

11.4.2 High solids potato

Studies have shown that increasing the ADPGPP activity through expression of the bacterial *glgC16* gene results in an increase in the rate of conversion of carbohydrate into starch, and that the overall level of starch accumulation is increased on average by 25%. Total ADPGPP activity is 3- to 4-fold higher in transgenic tubers, but the activities of other enzymes are not affected. These data show that a single enzymatic reaction can exert control over the level of accumulation of an important end product like starch. The data also show that manipulation of a single control point in a biochemical reaction can change the flux of metabolites through this pathway. In this case, changing the allosteric control properties of the ADPGPP enzyme results in an increased flow of carbon into starch. It is not the amount of enzyme *per se*, but the fact that the enzyme is not subject to normal metabolic control that is the key. This conclusion is supported by experiments where overexpression in potato tubers of bacterial ADPGPP enzymes which are subject to normal allosteric regulation did not result in any significant increase in starch content (Stark, 1992). Further experimental

Table 11.4 Sugar accumulation in high solids and control tubers stored at 3°C for 4 months. Reducing sugars are glucose and fructose, and total sugars are reducing plus sucrose. Values (fresh weight) represent the averages from 9 transgenic high solids lines and 11 control lines grown under growth-chamber conditions

	Reducing sugars	Total sugars	Starch
Transgenic	0.1	0.3	9.9
Control	0.8	1.0	6.0

evidence suggests that this enzymatic step, which normally controls the amount of starch produced, is no longer a controlling step in starch biosynthesis in the transgenic tubers expressing the deregulated bacterial ADPGPP enzyme (GlgC16) (Stark, 1992). This is based on the observation that no correlation exists between the level of gene expression and degree of starch increase once expression had reached a certain threshold level. This is not surprising since carbohydrate metabolism in general is a series of interdependent biochemical reactions with several different control points.

Perhaps as important as the increase in dry matter is the remarkable ability of transgenic tubers to withstand cold sweetening. Cold sweetening is the phenomenon where starch breaks down into sugars during low temperature storage of potatoes. Low temperature storage is desirable because low temperatures reduce shrinkage caused by respiration, losses due to sprouting, the need for chemical sprout inhibitors, and losses due to infection. However, there is a fairly low tolerance for sugars in the potato processing industry because sugars cause unacceptably dark pigments to form during processing. Transgenic high solids potatoes are able to withstand low temperatures and not accumulate sugars as do normal potatoes (Table 11.4). Further studies have revealed that the enhanced starch biosynthesis which leads to higher solids in developing tubers also functions to reconvert sugars into starch during storage. This agrees with biochemical models where the difference between varieties with good and poor storage characteristics lies in how the tuber partitions excess carbohydrate between sugars and starch.

In the case of the transgenic high solids tubers, the ability to rapidly convert sugars into starch is maintained during storage, and this has been studied in detail (Figure 11.1) (H. Davies and D. Stark, in preparation). In

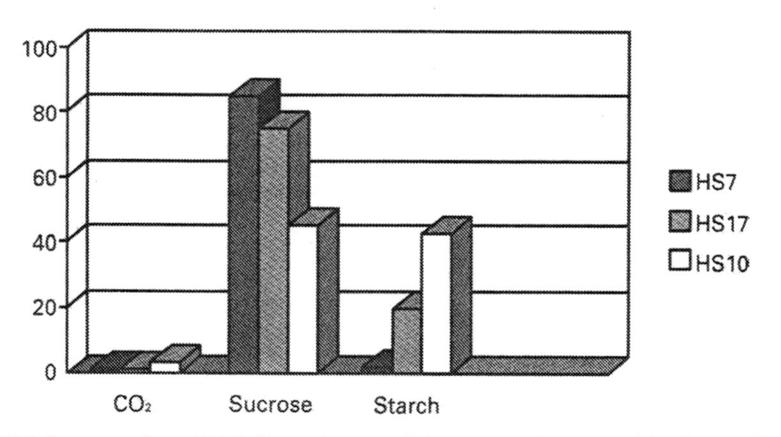

Fig. 11.1 Incorporation of 14C-glucose into starch in mature tubers stored 14 days at 10°C. Values indicate the % label recovered in CO_2, sucrose or starch in tubers of control line HS7 or transgenic lines HS10 and HS17.

these studies, radiolabeled sugars were injected into stored tubers and the incorporation into starch was measured. It was found that a normal Russet Burbank tuber had very little capacity to convert sugars into starch, while the transgenic high solids tubers maintained a fairly high capacity for starch biosynthesis even after 150 days of storage.

One way that sugars are removed prior to processing is to increase the storage temperature in a process called reconditioning. This stimulates metabolic activity in the tuber and results in sugars being removed via respiration and renewed starch biosynthesis. Reconditioning can take from a few days to a few weeks, and does not always result in potatoes with sufficiently low sugar levels for direct processing. In the transgenic high solids tubers, the sugars that do accumulate during storage are very rapidly converted into starch when the tubers are warmed to 16°C. The transgenic tubers typically produced light colored fries (an indication of low sugars) within as little as 24 hours at 16°C, where conventional potatoes would take 2 weeks or more during which time they begin to sprout and may require chemical sprout inhibitors (Table 11.5). Therefore the high solids potatoes represent a tool whereby storage management would be greatly improved and simplified relative to current practice. In summary, the bacterial ADPGPP gene imparts higher solids, lower sugar accumulation during low temperature storage and rapid metabolism of these sugars at warmer temperatures.

As mentioned previously, the one fatal flaw of the high solids technology is increased susceptibility to blackspot bruise. In general, the higher the solids content within a given variety, the greater the suceptibility to bruise damage. Bruise occurs when cells are ruptured, typically by physical impact during shipping and handling. The rupture of cellular membranes allows the enzyme polyphenol oxidase (PPO) to come in contact with its various phenolic substrates, typically amino acids such as tyrosine, leading to the formation of the dark pigment melanin. The bruise itself does not result in

Table 11.5 Reconditioning – Fry color

	Number of days at 16°C			
	0	3	6	10
Control	2.2	2.3	1.8	2.0
Control	3.5	3.3	2.5	1.7
High solids – 1	1.3	1.3	0.4	0.8
High solids – 2	3.2	1.0	0.5	1.0
High solids – 3	2.2	1.2	0.7	1.0
High solids – 4	1.0	0.7	0.5	0.5

Fry Color/Reconditioning response of high solids versus control potatoes. Fry color is rated according to the USDA chart on a scale of 0–4; lowest–highest. Potatoes were field grown and stored for 3 months at 4°C, and moved for the indicated number of days to 16°C. Fry colors below 2.0 are acceptable a change in flavor or nutritional properties, but is a major source of loss in the industry because the bruised tissue is not acceptable to consumers. In the transgenic high solids potatoes bruise susceptibility is increased for two reasons. First, there are dramatically more starch granules under the skin of a high solids potato which makes the tubers more susceptible to physical damage. These large, crystalline granules readily puncture membranes upon impact. Second, it was found that starch is not the only component of the tuber increased as a result of bacterial ADPGPP expression. In fact, all components of the tuber are increased indicating that total 'sink' activity, or the ability of the tuber to draw nutrients from the rest of the plant, is enhanced in the transgenic tubers. While this is scientifically of interest and supports several published hypotheses, the net effect is a 2-fold increase in the free amino acid pool. Therefore bruise susceptibility is greater due to an increased physical susceptibility to mechanical damage and to a 2-fold increase in tyrosine, the primary substrate for PPO.

While increased bruise susceptibility is a major issue hindering commercialization of the high solids technology, recently it has been shown that PPO activity can be reduced through expression of antisense RNA (Bachem, 1994). The reduction in PPO activity results in decreased incidence and severity of blackspot bruise. It will be of interest to determine if reduction of PPO activity via antisense RNA will reduce bruise susceptibility in the transgenic high solids tubers. Such experiments are in progress.

11.5 Health, safety and environmental issues

11.5.1 NewLeaf™ potato

Since the NewLeaf™ potato represents a transgenic food crop containing a pesticidal protein, approval was necessary prior to commercialization from all three relevant US regulatory agencies (FDA, USDA and EPA; see also Chapter 4). The USDA regulates the impact of genetically modified plants on the environment and production in agriculture under the jurisdiction of the plant Pest Quarantine Act. Essentially, the USDA ensures that the transgenic plant does not pose a threat to normal production agriculture by examining issues such as outcrossing, agronomic performance and impact on current agronomic practices. Through extensive field evaluations it was concluded that NewLeaf™ potatoes perform as any other Russet Burbank potato, with the exception of providing outstanding insect control (11.4, above). Unlike the EPA and FDA which primarily regulate the introduced proteins, each new variety of potato that is released with the *B.t.t.* gene must be approved by the USDA. Already, the varieties NewLeaf™ Atlantic and Superior have been registered and approved for commercial sale, and many other varieties will follow. The USDA retains its right to approve each new variety since each represents a new clonal selection and therefore acceptable agronomic performance and safety must be demonstrated prior to release.

The US EPA regulates environmental and food safety of pesticides under the jurisdiction of the Federal Insecticide, Fungicide and Rodenticide Act. The Food and Drug Administration oversees the safety and wholesomeness of food and feed under the Federal Food, Drug and Cosmetic Act. To satisfy these agencies, the published guidelines of 'substantial equivalence' were followed (see Chapter 4). For NewLeaf™, the safety of the introduced proteins, nutritional quality, level of important natural products and agronomic and environmental performance were all thoroughly measured (Lavrik, 1995). First, the *B.t.t.* protein was shown to be identical in amino acid sequence to that found in the naturally occurring microbe and to that found in commercial microbial formulations. This demonstrates that the same protein is found in the potato as that used to control insects for over 30 years. The safety of the protein was measured using acute gavage studies where animals were fed the equivalent of 2.5 million times the levels of *B.t.t.* protein that the average human would consume. This study and a simulated digestive fate study demonstrated that the *B.t.t.* protein caused no adverse effects in animals and is very quickly digested (less than 30 seconds) in the human digestive tract. Other studies included an allergenic assessment, insect specificity and environmental persistence, all of which showed that the new protein posed no risk to people, other insects or the environment. Compositional analyses for total protein, fat, carbohydrate and dietary fiber, and measurements of key vitamins, minerals and the natural toxicants found in potato (glycoalkaloids) showed that NewLeaf™ was substantially equivalent to Russet Burbank potatoes.

A similar safety assessment was carried out on the other introduced protein, NPT II. As with *B.t.t.*, it was concluded that the NPT II protein posed no risk to humans, animals or the environment. NPT II had been previously registered by the FDA as a food additive as part of the overall registration of Calgene's Flavr Savr™ tomato.

11.5.2 High solids potato

The type of starch produced in the high solids tubers is identical to normal potato starch, so there should be no issues of digestibility or food safety due to the presence of a novel starch.

11.6 Legislative and labeling position

11.6.1 NewLeaf™ potato

As mentioned above, the NewLeaf™ potato represents the first transgenic potato to be approved in the USA. Approval has also been granted in

Canada, Mexico, Japan, and Turkey, and is being sought in a number of Central and Eastern European countries. For import of frozen processed product into Western Europe, Novel Food Approval is being sought in the United Kingdom. Other approvals in Western Europe will be sought once the regulatory requirements are clarified. NewLeaf™ potatoes do not require segregation or labeling where approval has been granted. Public reception of the product has been so great that a portion of the crop is being offered under the NatureMark™ brand. It is important to note the distinction between branding and labeling. Labeling is in response to legal requirements to communicate important information to the consumer. Branding, in this case, is where a portion of the potato crop is set aside voluntarily, identity preserved and sold under a given name. This is similar to the marketing strategy used by Calgene to promote its Flavr Savr™ tomato (section 11.7).

11.6.2 High solids potato

Regulatory approval has not been initiated for the high solids potato.

11.7 Consumer acceptance and marketing

11.7.1 NewLeaf™ potato

The original intent was to market NewLeaf™ as any other commodity potato with no identity preservation. However, because this was one of the first transgenic products to reach the market, several end users, particularly in the quick service industry, showed some concern over consumer reaction to biotechnology-derived potatoes. Previous experience with Calgene's Flavr Savr™ tomato and Zeneca's California Purée, both advertised as products of biotechnology, suggested that consumer response would not be negative and actually may be positive if the added trait is perceived as being of benefit to the consumer or the environment. To gauge consumer reaction, potatoes were branded under the NatureMark™ name and clearly identified on the bag as being a 'genetically modified food'. The potatoes were offered in 10 lb bags in two different six-week test market studies, one in the mid-Atlantic region of the United States and the other in Atlantic Canada. In the United States, where the test market release was unpublicized and unpromoted, the release of genetically modified potatoes was 'uneventful'. Sales were equivalent to those of other potato offerings.

In Atlantic Canada, the test market was promoted through extensive news coverage, both television and newspaper, and through in-store displays. The potatoes were not otherwise advertised. Consumers responded to the environmental benefit of NewLeaf™ by purchasing 3 times as many 'genetically modified' potatoes as either the store brand or organically grown potatoes. Again, there was no negative response or backlash by the consumer or by groups of activists. This particular test was so positive that NatureMark™ branded potatoes have been requested by several different retailers. The concept of a branded potato is under further test markets in 1997.

11.7.2 High solids potato

The potential consumer benefits of a high solids potato are significant. The potatoes fry into a better tasting, crisper product with less oil and thus fewer calories from fat. The consensus feedback on this product concept is 'this is a great contribution from biotechnology' and 'when will the product be available?'.

11.8 Future prospects

NewLeaf™ potatoes are already being improved further with genes for conferring resistance to major potato viruses. Commercial launch of the first of these beetle and virus resistant plants will be in 1998. The environmental impact of these new varieties will be huge and will dramatically reduce or even totally eliminate the need for insecticide applications. These potatoes will eventually be made available in much of the world.

The number one threat to potato production worldwide is the disease called late blight, caused by the fungus *Phytophthora infestans*. Late blight was the disease that caused the Irish potato famine in the mid 1800s and is a re-emerging problem due to the occurrence of a new mating type of the fungus. This mating type allows the fungus to sexually propagate, enabling rapid adaptation to new environments and the development of resistance to chemical fungicides. In the United States in 1995, average costs to control the disease approached $200 per acre and the disease is still causing crop losses and reduced yields. Many research laboratories around the world are working on possible solutions to this devastating disease, and it is reasonable to predict that a NewLeaf™ potato with beetle, virus and late blight control will someday be available to growers.

A second important fungal disease is early dying caused by *Verticillium dahliae*. Early dying robs the grower of yield by causing early senescence and death of the vine. It is thought that early dying is a problem on most fields where potato is part of a normal crop rotation, and results in a 10–20% loss in yield. Monsanto researchers have found a solution to this disease that holds commercial promise. Under severe disease pressure, the transgenic fungal resistant potato plants showed extremely high levels of resistance against early dying and yield loss due to this disease. The beetle, virus and early dying resistance traits are being combined into a new generation of NewLeaf™ products and these should be available around the year 2002. Eventually, it is envisioned that growers will have a choice of controlling major pests through chemical means or by using built-in genetic control. Even fertilizers may eventually be reduced or even replaced by technology currently in development.

In the long run biotechnology promises to deliver a number of exciting and valuable products. Already, technology is being developed to make fried potato products more nutritious by using novel, healthful oils, increasing certain beneficial dietary fibers and increasing vitamin content. Ultimately, the huge productivity of potato as a crop may be exploited for production of renewable industrial raw materials such as plastics, pharmaceuticals and renewable sources of energy. While these products seem far off, in reality technical advancements made in laboratories around the world are showing that these goals are in the not-so-distant future.

References

Bachem, C. W. B. *et al.* (1994) Antisense expression of polyphenol oxidase genes inhibits enzymatic browning in potato tubers. *Bio/Technology* 12, 1101–5.

Cattaneo, J. *et al.* (1969) Genetic studies of *Escherichia coli* K12 mutants with alterations in glycogenesis and properties of an altered adenosine-diphosphate-glucose-pyrophosphorylase. *Biochem. Biophys. Res. Commun.* 34, 694–701.

Klee, H. *et al.* (1987) Agrobacterium-mediated plant transformation and its further application to plant biology. *Annu. Rev. Plant Physiol.* 38, 467–86.

Kuziel, M. G. *et al.* (1993) The insecticidal crystal proteins of *Bacillus thuringiensis*: past, present and future uses. *Biotechnol. and Genet. Engineer. Rev.* 11, 171–228.

Lavrik, P. B. *et al.* (1995) Safety assessment of potatoes resistant to Colorado potato beetle, in *Genetically Modified Foods* (eds K. -H. Engel, G.R. Takeoka and R. Teranishi), American Chemical Society symposium series 605, pp. 148–58.

Lee, Y. M., Kumar, A. and Preiss, J. (1987) Amino acid sequence of an *Escherichia coli* ADPglucose synthetase allosteric mutant as deduced from the DNA sequence of the *glgC* gene. *Nucleic Acids Res.* 15, 10603.

Perlak, F. J. *et al.* (1993) Genetically improved potatoes: protection from damage by Colorado potato beetles. *Plant Mol. Biol.* 22, 313–21.

Preiss, J. (1991) Biology and molecular biology of starch synthesis and its regulation, in *Oxford Surveys of Plant Molecular and Cell Biology*, Vol. 7. (ed. B. Miflin), Oxford Press, pp 59–114.

Stark, D. M. *et al.* (1992) Regulation of the amount of starch in plant tissues by ADPglucose pyrophosphorylase. *Science* 258, 287–92.

12 Cereals

PAUL CHRISTOU

12.1 Introduction and historical perspective

The engineering of cereal crops has become possible only relatively recently as a result of the development of particle-bombardment based methodology. The cereals are monocotyledons and exhibit natural resistance to infection by *Agrobacterium tumefaciens*, the vector commonly used to transform dicotyledons. Even in situations in which such resistance had been overcome, a variety of problems has prevented the wider adoption of *Agrobacterium*-based vectors in monocotyledons.

One of the advantages of particle bombardment is the capability of introducing biologically functional DNA into organized and regenerable tissues, opening the way to the more efficient engineering of cereals. An additional factor which has contributed significantly to the development of effective gene transfer methods for cereals has been the refinement of ubiquitin promoter-based vectors (Christensen and Quail, 1996). Engineering of maize, wheat and barley now relies exclusively on this promoter because promoters which are active in dicotyledonous species are not expressed in cereals other than rice.

In this chapter, attention is focused on the four major cereals: maize, wheat, rice and barley. Major agrochemical and breeding companies have invested substantial resources in terms of labour, time and cash to bring effective gene transfer systems on line for maize and wheat, and to a lesser extent for barley. Rice, being a non-traded commodity, has attracted significantly less attention from the commercial sector; however, international organizations such as the Rockefeller Foundation, have spearheaded a comprehensive effort to make the tools of plant molecular biology available to rice researchers in the western world and more importantly in rice growing countries in the tropics. Furthermore, the International Agricultural Research Centers (IARCs) specialize in particular crops and provide a comprehensive research basis for their improvement with emphasis on developing countries. For example, the International Rice Research Institute (IRRI, Los Banos, The Philippines) has the mandate for rice improvement in Asia. Wheat and maize research is conducted at the Centro International de Mejoramiento de Maize y Trigo (CIMMYT). CIAT

Table 12.1 International organizations facilitating the transfer of plant biotechnologies to developing countries

International organization*	Description
Food and Agriculture Organization (FAO) of the United Nations, Rome, Italy	Conducts research on and facilitates transfer of plant biotechnologies that can benefit developing countries through its Plant Production and Protection Division in Rome, its regional offices and the Joint FAO/IAEA Division of Nuclear Techniques in Food and Agriculture in Vienna.
International Atomic Energy Agency (IAEA), Vienna, Austria	Conducts collaborative research with national agencies and provides training in mutation breeding and other plant biotechnologies through the Joint FAO/IAEA Division of Nuclear Techniques in Food and Agriculture.
International Laboratory for Tropical Agricultural Biotechnology (ILTAB), La Jolla, CA, USA	An advanced-research laboratory, developed through a collaboration between the Scripps Research Institute and the French technical assistance organization ORSTOM, which conducts research and offers training on development of disease resistant tropical plants through genetic engineering.
International Center for Genetic Engineering and Biotechnology (ICGEB), Trieste, Italy and New Delhi, India	Originally established by the United Nations Industrial Development Organization and now an independent research and training organization with crop biotechnology programs in New Delhi and information dissemination provided through Trieste.
Center for the Application of Molecular Biology to International Agriculture (CAMBIA), Canberra, Australia	A research and technology transfer organization specializing in the production and dissemination of inexpensive biotechnology tools that can be employed in developing countries.
International Service for the Acquisition of Agri-biotech Applications (ISAAA), Ithaca, NY, USA	An international organization committed to the acquisition and transfer of proprietary agricultural biotechnologies from the industrial countries for the benefit of the developing world.
Intermediary Biotechnology Service (IBS), The Hague, The Netherlands	A unit of the International Service for National Agricultural Research which provides national agricultural research agencies with information, advice, and assistance to help strengthen their biotechnology capacities.
Biotechnology Advisory Commission (BAC), Stockholm, Sweden	A unit of the Stockholm Environment Institute that provides biosafety advice and helps developing countries assess the possible environmental, health, and socioeconomic impacts of proposed biotechnology introductions.
Technical Center for Agricultural and Rural Development (CTA), Wageningen, The Netherlands	A unit of the European Union that collects, disseminates and facilitates exchange of information on research innovations including plant biotechnologies for the benefit of Asian, Caribbean and Pacific States.
International Institute for Co-operation in Agriculture (IICA), San José, Costa Rica	Assists countries in Latin America and the Caribbean with policy issues related to biotechnology including the formulation and harmonization of biosafety procedures.

* International organizations that are primarily funders are not included in this list.
Reprinted from Toenniessen (1995) *Trends in Biotechnology* 13: (9) p. 406.

(Centro Internacional de Agricultura Tropica) and WARDA (West Africa Rice Development Association) have the mandate for the development of rice varieties suitable for cultivation in South America and West Africa, respectively. Barley research is conducted primarily at ICARDA (International Center for Agricultural Research in Dry Areas) in Syria where research on wheat is also being carried out.

A number of international organizations have been set up to facilitate the transfer of plant biotechnology to developing countries, as shown in Table 12.1 (Toenniessen, 1995). Toenniessen has also developed a model which illustrates various facets of the international agricultural system, in which cereals feature very prominently. He has also demonstrated elegantly how these organizations fit into a continuum, beginning with fundamental research, which also includes, in sequence, strategic and applied components, leading to adaptive and finally operational research activities (Figure 12.1).

12.2 The techniques of genetic modification in cereals

The most widely used method for the engineering of cereals is now particle bombardment (Klein *et al.*, 1987). The technique, reviewed extensively

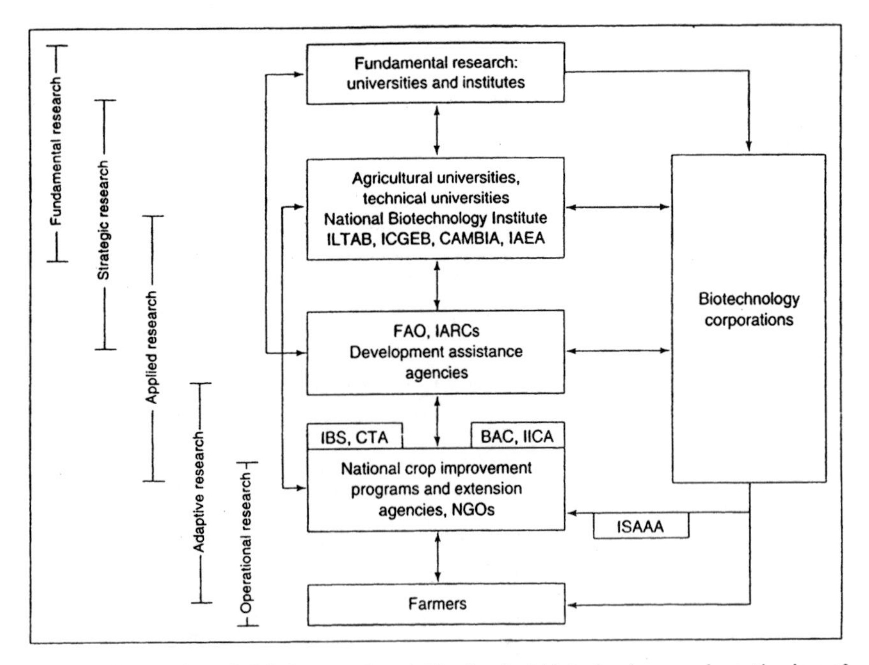

Fig. 12.1 Integration of global research activities in plant biotechnology – schematic view of the biotechnology international agricultural research system. (Reprinted from Toenniessen (1995) *Trends in Biotechnology* 13 (9) p. 407, with permission.)

elsewhere (Yang and Christou, 1994; Christou, 1993), is based on the concept of accelerating DNA-coated metal particles into tissues from which intact transgenic plants may be regenerated, ideally after a selection step. The concurrent development of culturing systems compatible with bombardment technology has also been vital for the successful engineering of cereals.

12.2.1 Maize

Research activity in maize biotechnology started in commercial laboratories, primarily in the USA and to a lesser extent in Europe (Pioneer HiBred, AgrEvo, Monsanto, DeKalb Genetics, Ciba-Geigy, Zeneca, Sandoz, Plant Genetic Systems and others). Early work with electroporated transformed protoplasts resulted in major disappointment when all transgenic plants obtained were shown to be sterile (Rhodes *et al.*, 1988). Furthermore, although *Agrobacterium* was effective in introducing viral DNA into maize cells, it was incapable of transforming the genome (Grimsley *et al.*, 1987).

In 1990, two groups working independently achieved the successful recovery of fertile transgenic plants using particle bombardment of embryogenic maize suspension cultures and selection using the herbicide bialaphos (Fromm *et al.*, 1990; Gordon-Kamm *et al.*, 1990). Stable inheritance and expression of *bar*, the gene responsible for encoding the enzyme phosphinothricin acetyltransferase (PTT) which, in turn, confers resistance to bialaphos, were demonstrated in a number of subsequent generations. The transformation process and the presence of a foreign gene did not affect plant vigour or fertility (Gordon-Kamm *et al.*, 1991).

Subsequently, alternatives to particle bombardment, such as electroporation of wounded immature embryos or embryogenic callus, were also developed (D'Halluin *et al.*, 1992). Tissues were wounded either mechanically or enzymatically and electroporated with a neomycin phosphotransferase II gene. Transformed embryogenic calli were selected on kanamycin-containing medium and fertile transgenic plants were successfully regenerated.

In spite of the developments described above, it is still not possible to produce regenerable cultures from many of the elite maize varieties used commercially. Consequently, further refinements of techniques are necessary to make the genetic engineering of maize effective and practical. For example, explants commonly used for maize transformation via bombardment include immature embryos, as well as callus or suspension cultures derived from immature embryos. These methods are time-consuming and labour-intensive and there is a need to maintain donor plants in the greenhouse year-round as sources of material for excisions.

To develop a less genotype-dependent maize transformation procedure, Wan *et al.* (1995) used 10-month-old callus as target tissue for microprojectile bombardment. Twelve transgenic callus lines were obtained from two of the three anther culture-derived callus cultures representing different genetic backgrounds. Multiple fertile transgenic plants (R0) were regenerated from each transgenic callus line. The introduced genes were successfully integrated into the plant genome and expressed in the new transgenic lines as resistance to the herbicide Basta®. This study suggested that maize callus could be transformed efficiently through microprojectile bombardment, that fertile transgenic plants could be recovered and that the system should facilitate the direct introduction of agronomically important genes into commercial genotypes.

In an attempt to develop a facile and variety-independent gene transfer system for maize, *in vitro* methods have been further optimized to regenerate clumps of multiple shoots and somatic embryos at high frequency from shoot tips of aseptically-grown seedlings as well as from shoot apices of germinated immature embryos (Heng *et al.*, 1992). The shoots were rooted easily and transferred to the greenhouse where they grew into normal plants. The sweet-corn genotype, Honey N Pearl, was used for these experiments, but shoot-tip cultures from all of 19 other corn genotypes tested also responded in a similar manner.

Based on the above procedure, Zhong *et al.* (1996) developed a reproducible system for the recovery of fertile transgenic maize plants. The transformation was carried out by bombarding cultured shoot apices with a plasmid carrying the *bar* gene and the potato proteinase inhibitor II gene (*pin2*, Kiel *et al.*, 1986). Bombarded shoot apices were multiplied in culture and selected with glufosinate ammonium. Co-transformation frequency was 100% for linked genes and 80% for unlinked genes. This system has distinct advantages over alternative culture systems for transformation. Shoot tips have been propagated in culture for a very long time (more than four years) without loss of regeneration potential. In order to assess expression of the *bar* gene, the herbicide Ignite® was sprayed on all recovered plants. All engineered plants were shown to be resistant to the herbicide, whereas untransformed control plants exhibited necrosis in two days and died 10 days after application of the herbicide. Using this system, the authors reported recovery of regenerated plantlets from 36 different genotypes.

Ishida *et al.* (1996) have recently described a transformation system for maize mediated by *A. tumefaciens*. Transformants of the inbred line A188 were recovered following cocultivation of immature embryos with a 'superbinary vector'. This is a significant advance in the field, however, it will be important to demonstrate transformation of elite maize germplasm using this method before it can be incorporated into routine programmes for the improvement of maize.

12.2.2 Wheat

Particle bombardment has again provided the necessary technological breakthrough for engineering wheat. A plasmid encoding *bar* (conferring resistance to the herbicide bialaphos) has been introduced into type C long-term regenerable embryogenic callus (Vasil *et al.*, 1992). Following selection on bialaphos-containing medium, transgenic plants from two lines were regenerated. Transgenic progeny was obtained by crossing the transgenic primary transformants with wild-type pollen. Subsequently, the same group reported an improved procedure in which DNA was bombarded directly into the scutellum of immature embryos of two spring and one winter cultivar of wheat (Vasil *et al.*, 1993). Twelve independent callus lines showing PPT activity were recovered from the bombardment of 544 explants (either immature embryos or calli derived from them). R0 plants were regenerated from seven of these lines, of which five produced seed. Resistance to topical applications of the herbicide bialaphos was confirmed. Incremental improvements in the procedure were reported later, however, most groups involved in wheat genetic engineering still focus on one or two specific varieties which exhibit a prolific regeneration response in culture.

Weeks *et al.* (1993) utilized the cultivar Bobwhite because of its high frequency of regeneration from tissue culture. Immature embryos were bombarded with a plasmid containing the *bar* and *gusA* genes driven by the maize ubiquitin promoter. Nine independent fertile transgenic plants were obtained at a transformation frequency of approximately 0.1–0.2%. Some of the lines had normal fertility and seed set; however, most exhibited reduced seed set compared with wild type plants regenerated from tissue culture; one line was completely sterile. Amongst the various criteria used to assess transformation, the ability of regenerated plantlets to develop a normal root system in the presence of bialaphos was the most reliable indication of stable integration of the *bar* gene. Similar procedures, with additional refinements, were used by Nehra *et al.* (1994) and Becker *et al.* (1994).

As an alternative to particle bombardment, transformation of protoplasts from an embryogenic suspension culture of wheat (cv Hartog) using electroporation has also resulted in the recovery of a number of transgenic plants expressing *bar* and *gusA* (He *et al.*, 1994). However, no progeny analysis was reported in these experiments, so it is difficult to assess how useful this procedure will be.

The lack of alternative selectable markers in crop transformation has been a substantial barrier for commercial application of agricultural biotechnology. Zhou *et al.* (1995) have developed an efficient selection system for wheat transformation using glyphosate-tolerant CP4 and GOX genes as selectable markers. Immature embryos of the wheat cultivar Bobwhite were bombarded with two separate plasmids harbouring the

CP4/GOX and *gusA* genes. After one week, the bombarded embryos were transferred to a selection medium containing glyphosate. Transgenic plants tolerant to glyphosate were successfully recovered. Southern blot analysis demonstrated that the transgenes were integrated into the wheat genomes and transmitted to the following generation. The use of CP4 and GOX genes as selectable markers now provides an effective alternative transformation selection system for wheat.

Efficient procedures for the engineering of a number of different varieties of wheat including elite commercial varieties are rumoured to be in place in several companies including Ciba-Geigy, Nickersons and others (personal communications).

12.2.3 Rice

Amongst the four major cereals, rice has been by far the easiest to manipulate. Regeneration from protoplasts has been accomplished and is routine for at least a few japonica varieties. This plasticity has enabled many laboratories to launch extensive programmes on various aspects of rice biotechnology. The spectacular successes that have been accomplished in rice genetic engineering are due primarily to the establishment and implementation of the Rice Biotechnology Program funded by the Rockefeller Foundation (Anon., 1995).

The recovery of transgenic rice plants was first reported in the late 1980s. These experiments utilized direct DNA transfer methods such as electroporation (e.g. Toriyama *et al.*, 1988) or polyethylene glycol (PEG)-mediated gene transfer (e.g. Zhang *et al.*, 1988) into protoplasts. Subsequent reports established these two techniques as the methods of choice for gene transfer in rice. A severe limitation of these methods, however, is that most elite japonicas as well as the vast majority of indica varieties are very difficult to regenerate from protoplasts. Exceptionally, Datta and co-workers have reported the successful recovery of fertile transgenic plants from the variety Chinsurah Boro II (Datta *et al.*, 1990). Later, primary transformants from the elite variety IR72 were also reported by the same group; however, these were infertile (Datta *et al.*, 1992). In both cases, embryogenic suspension cultures were established from immature pollen grains and these cultures were used to isolate protoplasts for subsequent transformation experiments. These results highlighted some of the major problems with protoplast-based systems including the strong genotype- and culture-dependency of rice *in vitro* culture and reduced fertility or complete infertility of primary transformants.

As with the other cereals described in this chapter, particle bombardment technology provided the necessary breakthrough for the efficient genetic transformation of rice. The creation of genetically engineered rice from every variety is now possible. Following particle bombardment, Christou *et al.* (1991), recovered transformed plants at very high frequencies in the range of 2 to 15% calculated in terms of independently-derived transgenic plants per unit number of bombarded explants. The antibiotic hygromycin B was used to select transformed embryogenic callus. All recovered plants were clonal in nature, demonstrating their derivation from single or a small number of cells (Christou *et al.*, 1992). Transfer of embryogenic callus to appropriate media following a secondary selection (Christou and Ford, 1995) resulted in the development of plants expressing marker genes, in addition to antibiotic/herbicide-resistance genes. Stable transformation in R0 plants and their progeny was confirmed by extensive molecular, biochemical and genetic analyses. Field trials in which resistance to the herbicide Basta® was demonstrated, were carried out in 1993 (see later sections).

Cao *et al.* (1992) used suspension culture cells to generate herbicide-resistant plants expressing *bar*. They used the herbicide as a selectable marker confirming earlier results that phosphinothricin can be as effective as hygromycin for the selection of transformed cells. Zhang *et al.* (1996) and Sivamani *et al.* (1996) also developed regenerable embryogenic suspension cultures of elite indica rice varieties which resulted in the recovery of transgenic plants expressing marker genes. An advantage of using suspension cultures is the ability to obtain material for transformation experiments derived from mature seed, thus eliminating the requirement for growing donor plants as source material. Disadvantages of the procedure, however, are the requirement for labour-intensive maintenance of embryogenic suspension cultures, and the loss of regeneration potential of such cultures upon prolonged culture periods.

Meanwhile, Hiei *et al.* have reported an elegant study in which *A. tumefaciens* was re-engineered to allow infection of rice cells in culture. High efficiencies of transformation were reported with two elite japonica varieties. Other investigators have now been able to reproduce *Agrobacterium*-mediated transformation of rice. It is now possible, therefore, to compare bombardment- and *Agrobacterium*-based transformation methods for rice.

12.2.4 Barley

Wan and Lemaux (1994) have described a rapid, efficient, and reproducible system to generate large numbers of transgenic barley (*Hordeum vulgare* L.) plants. Immature embryos, young callus, and microspore-derived embryos were bombarded with a plasmid containing *bar* and *gusA* either alone, or in combination with another plasmid containing a barley yellow dwarf virus coat protein (BYDVcp) gene. A total of 91 independent bialaphos-resistant callus lines expressed functional phosphinothricin acetyltransferase, the product of *bar*. Integration of *bar* was confirmed by DNA hybridization. Cotransformation frequencies of 84 and 85% were determined for the two linked genes (*bar* and *gusA*) and for two unlinked genes (*bar* and the BYDVcp gene) respectively. More than 500 green, fertile, transgenic plants were regenerated from 36 transformed callus lines on bialaphos-containing medium. Transmission of the genes to R1 progeny was confirmed by DNA hybridization. A germination test of immature R1 embryos on bialaphos-containing medium was useful for selecting individuals that were actively expressing *bar*, although this was not a good indicator of the presence or absence of *bar*. Expression of the gene in some progeny plants was indicated by resistance to the herbicide Basta®. The R1 plants were in soil approximately seven months after bombardment of the immature embryo.

Transgenic, fertile barley plants from the Finnish elite cultivar Kymppi have been obtained by particle bombardment of immature embryos and without selection (Ritala *et al.*, 1994). One out of a total of 227 plants expressed the transferred *nptII* gene. Fertile transgenic barley plants have also been obtained using the plasmid pBC1 to deliver the selectable *hph* gene and reporter *gusA* gene into immature embryos (Hagio *et al.*, 1995). Following selection, 18 hygromycin resistant plants were obtained. Transmission of the *hph* gene to progeny (R1) of two independent R0 plants was confirmed by Southern hybridization. Transformation frequencies with this system were generally lower than other published procedures; however, this work is significant as it demonstrates that the antibiotic hygromycin could be employed effectively for selection of transgenic barley.

Jahne *et al.* (1994) have developed a system for the bombardment-mediated transformation of barley using freshly isolated microspores as the target tissue. Molecular and genetic analyses confirmed stable integration of *bar* and *gusA* genes into progeny. The procedure, however, was effective on only one variety of barley which regenerates effectively from microspores. A procedure utilizing transformation of protoplasts from the same variety also resulted in the recovery of transgenic plants and progeny (Funatsuki *et al.*, 1995). However both procedures will remain of limited value unless they can be extended to elite commercial varieties. The most efficient procedure reported for barley transformation remains that developed by Wan and Lemaux (1994), described above.

12.3 The application of gene technology in cereals

Previous sections illustrated the development of effective procedures for the introduction of genes into important cereal crops. In this section, some of the useful genes which have been introduced into crops are examined. The targets of applied biotechnology fall into three general categories: the modification of agronomic properties, the alteration of industrial traits, and the development of alternative uses for crop plants. It is difficult to paint an accurate picture of the status of R&D for these crops as most of the activity is carried out in commercial laboratories and is highly confidential. Where possible, examples are given and supplemented by information obtained through personal communication, patent applications, press releases and meeting abstracts.

Field trials with transgenic plants including only two cereals (maize and rice) have been ongoing for a number of years, primarily in the USA. Table 12.2 summarizes worldwide field trial activity in different crops over several years while Figure 4.2 in Chapter 4 shows the number of field trials notified in the USA alone.

12.3.1 Maize

Immediate targets in maize biotechnology include pest, insect and microbial resistances. Most of the research activities which have led to field trials and commercialization to date have focused on the generation of insect resistance utilizing *Bacillus thuringiensis* (Bt) genes. For example, the European corn borer (ECB; *Ostrinia nubilalis* Hubner) is an economically significant pest of corn and is sensitive to the Bt toxin. Koziel *et al.* (1993) have introduced a synthetic gene encoding a truncated version of the insecticidal protein derived from *B. thuringiensis* into an elite line of maize via particle bombardment. Hybrid maize plants resulting from crosses of transgenic elite inbred plants with commercial inbred lines were evaluated for resistance to the European corn borer under field conditions. Plants expressing

Table 12.2A Most frequently released transgenic crops in the USA, 1987 to April 1997

Crop	Number of permits or notifications
Maize	1236
Tomato	347
Soybean	303
Potato	298
Cotton	218
Melon & squash	116
Tobacco	104
Rapeseed	65
Sugarbeet	33
Alfalfa	19
Wheat	17
Cucumber	16
Rice	16
Sunflower	9
Creeping bentgrass	8
Lettuce	7
Apple	7
Peanut	5

Source: USDA/APHIS

Table 12.2B Field releases of transgenic cereals in the USA from 1987 to April 1997

Maize

Trait	First release and year	Subsequent releases
Marker genes	BioTechnica (1990)	Ciba-Geigy, Garst/ICI
Herbicide resistance	DeKalb (1991)	Ciba-Geigy, Upjohn Monsanto, Pioneer, Holdens, Pioneer, Cargill, Northrup King, Hoechst
Storage protein	BioTechnica (1991)	DeKalb
Insect resistance	Monsanto (1992)	Pioneer, Northrup King, Ciba Geigy, DeKalb
Virus resistance	Pioneer (1992)	Northrup King
Pigment metabolism	Pioneer (1992)	
Male sterility	Cargill (1992)	Holdens, Pioneer
Fungal resistance	Northrup King (1995)	Pioneer

Rice

Trait	First release and year	Subsequent releases
Transposons	Louisiana State University (1990)	
Marker genes	Pennsylvania State University (1990)	
Insect resistance	Louisiana State University (1991)	
Seed composition	Louisiana State University (1992)	
Herbicide resistance	Louisiana State University (1993)	AgrEvo
Fungal resistance	Louisiana State University (1995)	
Bacterial resistance	Univ of California (1996)	
Pharmaceutical proteins	Applied Phytologicals Inc. (1996)	

Wheat

Trait	First release and year	Subsequent releases
Herbicide resistance	Monsanto (1994); AgrEvo (1994)	
Altered color	Monsanto (1994)	
Marker genes	Monsanto (1995)	
Fungal resistance	Monsanto (1995)	
Viral resistance	Monsanto (1995)	Univ. of Idaho
Storage protein, seed composition	US Agric. Res. Service (1995)	

Source: USDA/APHIS

high levels of the insecticidal protein exhibited excellent resistance to repeated heavy infestations of the insect.

Similarly, synthetic versions of insecticidal protein genes from *B. thuringiensis* subsp. *kurstaki* (Btk) were introduced into the Hi-II (A188/B73 derivative) genotype of corn (Armstrong *et al.*, 1995). Of 715 independent transgenic calli produced, 314 (44%) had insecticidal activity

Table 12.2C Most frequent categories of traits introduced into transgenic crops in field trials in the USA, 1987 to April 1997

Trait	Actual number	Percentage
Herbicide tolerance	902	28.2
Viral resistance	334	10.4
Insect resistance	784	24.5
Other (marker genes, bacterial nematode resistances)	223	7.0
Product quality	827	25.8
Fungal resistance	130	4.1

Source: USDA/APHIS

against tobacco hornworm (*Manduca sexta* L.) larvae. Plants were regenerated, self-pollinated when possible, and crossed to B73. First-generation progeny were evaluated under field conditions with artificial ECB infestations in 1992/1993. Approximately half segregated in a single-gene manner for resistance to first-generation ECB leaf-feeding damage. All of the lines evaluated for resistance to second-generation ECB exhibited less stalk tunnelling damage than the non-transgenic controls. In 1993, 44% of the lines tested had less than or equal to 2.5 cm of tunnelling, compared to severe damage (mean = 45.7 cm) in the B73 × Hi-II controls.

Herbicide resistance has also been a major target with primary emphasis in Roundup (Monsanto) and glufosinate (AgrEvo) resistance. These herbicides offer significant advantages over pre-emergence herbicides, as resistant transgenic plants allow post-emergence applications as needed depending on the level, severity and timing of weed infestation. A number of additional field trials including herbicide resistance have been carried out successfully, including one by the company AgrEvo in which maize plants engineered for resistance to the herbicide glufosinate were evaluated successfully in the field (personal communication) and a series of trials by Plant Genetic Systems (Gent, Belgium) evaluating engineered male sterility in maize (personal communication).

Bacterial, virus and fungal resistances are all targets currently under investigation. Utilization of transgenic maize for the production of pharmaceutical and industrial enzymes, macromolecules, biodegradable plastics or modified starches are also targets that a number of research groups are concentrating on, particularly in industry. Improvements in the nutritional quality of protein are exemplified by a recent publication reporting lysine accumulation in maize cell cultures engineered with a lysine-insensitive form of maize dihydrodipicolinate synthase (Bittel *et al.*, 1996).

12.3.2 Wheat

Herbicide-resistant wheat has been generated by a number of researchers, as described in section 12.2. In addition, fungal resistance is a major target for wheat, utilizing glucanases and chitinases. The biotechnology of breadmaking is another target in wheat genetic engineering. The ability to improve dough-making quality of wheat flour depends largely on the visco-elastic properties conferred to wheat dough by the gluten proteins (Shewry *et al.*, 1995). One group of gluten proteins, which include the high molecular weight subunits, are largely responsible for elasticity. Variation in their amount and composition is associated with difference in elasticity and thus quality between various types of wheats. Direct transformation of wheat with genes able to alter these properties is an immediate and feasible target.

12.3.3 Rice

Because of the relative ease with which transgenic rice plants can be generated, many researchers have examined gene function and organization, as well as cell, tissue and organ-specific expression in transgenic rice. In this respect, rice has served as the model system for monocots, much the same way as Solanaceous plants and Arabidopsis have been for dicots. The major targets for rice biotechnology are summarized in Table 12.3. In the following section, specific examples which might be deployed in the field in the not-too-distant future (i.e. within the next 3–5 years) are discussed.

12.3.3.1 Herbicide resistance. A number of elite rice varieties engineered for resistance to the non-selective herbicide gluphosinate ammonium have been the subject of extensive field trials with expected commercialization before the end of the century (Oard *et al.*, 1996). Bialaphos-resistant rice plants have also been generated; these have been shown to be completely protected from infection by mycelia of *Rhizoctonia solani*, the organism that causes sheath blight, after treatment with the herbicide (Uchimiya *et al.*, 1993). Substantial suppression of the disease was also observed when bialaphos was applied to transgenic plants infected with *R. solani* two days

Table 12.3 Rice biotechnology targets

Disease resistances	Quality traits	Tolerance
Tungro virus	cytoplasmic male sterility	drought
Yellow stemborer	apomixis	submergence
Gall midge	nutritional improvement	lodging
Brown planthopper		waterlogging
Ragged stunt virus		cold
Leaffolder		saline conditions
Sheath blight		acid sulfate soil
Storage pests		
Bacterial blight		
Blast		
Striped stemborer		
Whitebacked planthopper		

before herbicide treatment. A number of factors need to be examined very carefully prior to establishing whether this will be a viable strategy for combating sheath blight in rice. Bialaphos is an expensive chemical and it is not known whether it will persist in the field following spraying. If persistence is not an issue in terms of combating the disease, it could be an issue from a regulatory standpoint. However, experiments described in this report are interesting and may lead to the development of alternative strategies for disease management.

The commercial cultivars Gulfmont, IR72 and Koshihikari have also been genetically modified to express the *bar* gene which confers resistance to the nonselective herbicide glufosinate (Christou *et al.*, 1991, 1992). A field study was conducted in 1993–1995 at the Rice Research Station near Crowley, Lousiana, in which transgenic and non-transgenic plants were treated at the 4-leaf stage with 1.12 or 2.24 kg/ha glufosinate (Oard *et al.*, 1996). All 11 independently-derived transgenic lines produced normal looking seed at maturity. Significant improvements in performance, particularly in terms of grain yield, were observed with the transgenic plants. These results demonstrated that the *bar* gene was effective in conferring field-level resistance to glufosinate in rice.

12.3.3.2 Bacterial resistance. The Xa21 gene, which confers resistance to *Xanthomonas oryzae pv. oryzae* race 6, has been isolated and introduced into transgenic rice plants which subsequently displayed high levels of resistance to the pathogen (Song *et al.*, 1995). Further characterization of Xa21 should facilitate understanding of plant disease resistance mechanisms and lead to engineered bacterial resistance in rice in the future.

A new basic chitinase gene, designated RC24, has been isolated from a rice genomic library (Xu *et al.*, 1996). The predicted RC24 protein contains 322 amino acid residues and exhibited 68% to 95%, amino acid identity with known class I rice chitinases. RC24 protein expressed in *Escherichia coli* exhibited chitinase activity and strongly inhibited bacterial growth. A basal level of RC24 transcripts was detected in rice root and stem tissues, but not in leaf tissues. RC24 transcripts rapidly accumulated within 1 h after fungal elicitor treatment of suspension-cultured cells, and the levels continued to increase for at least 9 h. RC24 transcript accumulation was also observed in intact leaf tissues upon wounding. Transgenic rice plants containing the RC24/GUS gene fusion further confirmed that the RC24 gene showed a tissue-specific expression pattern and that transcription of the RC24 promoter was sensitively and rapidly activated by wounding.

12.3.3.3 Insect resistance. Insect pests are important targets for rice biotechnology. In experiments targeted towards development of insect-resistant rice, Fujimoto *et al.* (1993) reconstructed the CrylA *Bt* gene and produced transgenic plants which efficiently expressed the modified gene.

Bioassays using R2 generation plants with two major rice insect pests, the striped stem borer (*Chilo suppressalis*) and the leaffolder (*Cnaphalocrosis medinalis*) indicated that transgenic plants expressing the CrylA(b) protein were more resistant to these pests than wild-type control plants. Feeding studies with transgenic rice plants from the indica breeding line IR58 engineered with a synthetic *CrylA(b)* gene from *Bacillus thuringiensis*, showed mortality of up to 100% for two of the most destructive lepidopterous insect pests of rice in Asia, the yellow stem borer (*Scirpophaga incertulas*) and the striped stem borer (*Chilo suppressalis:* Wunn *et al.*, 1996). Feeding inhibition of the two leaffolder species *Cnaphalocrocis medinalis* and *Marasmia patnalis* was also demonstrated. Even though transformation frequency was low, these experiments represent a significant contribution towards the development of tools for the engineering of insect resistance in important rice breeding lines.

Corn cystatin (CC), a phytocystatin, shows a wide inhibitory spectrum against various proteinases. Many insect pests, especially Coleoptera, have cysteine proteinases which probably act as digestive enzymes. Irie *et al.* (1996) have introduced the corn cystatin gene into rice as a first step towards obtaining a rice plant with insecticidal activity. The transgenic rice plants contained high levels of corn cystatin in both seeds and leaves, reaching a level of up to 2% of the total heat soluble protein in seed. Corn cystatin prepared from transgenic rice plants showed potent inhibitory activity against proteinases that occur in the gut of the insect pest *Sitophilus zeamais*.

12.3.3.4 Virus resistance. The coat protein gene of rice stripe virus has been introduced into two japonica varieties (Hayakawa *et al.*, 1992). Transgenic plants expressed the coat protein at high levels and exhibited a significant level of resistance to virus infection. Plants derived from progeny of the primary transformants also expressed the coat protein and showed resistance to the virus, indicating stable transmission of the gene to the next generation. The virally-encoded stripe disease-specific protein was not detected in transgenic plants expressing the coat protein eight weeks after inoculation, demonstrating protection prior to viral multiplication.

12.3.3.5 The waxy *gene.* Shimada *et al.* (1993) have reported experiments in which a portion of the sequence of the rice *waxy* gene encoding a granule-bound starch synthase was inserted in an antisense orientation into rice plants. Seeds from some of the regenerated plants showed a significant reduction in the amylose content of grain starch. This study showed that antisense constructs can result in a reduced level of expression of the target gene.

12.3.3.6 Seed storage proteins. The seed storage protein beta-phaseolin of the common bean (*Phaseolus vulgaris* L.) has been expressed in transgenic rice plants (Zheng *et al.*, 1995). The highest quantity of phaseolin detected was 4.0% of the total endosperm protein in the transgenic rice seeds. The phaseolin trait was stably inherited through three successive generations. Immunolabelling studies using light and electron microscopy demonstrated that phaseolin accumulated primarily in the vacuolar protein bodies located at the periphery of the endosperm near the aleurone layer of the rice grain. These results have strong implications for the nutritional improvement of rice in the future.

12.3.4 Barley

The creation of herbicide-resistant barley has been described in section 12.2.4. Research directed towards improving the malting qualities of barley is being pursued primarily in industrial laboratories. Due to intense competition in the brewing industry, very little has been disclosed in the scientific literature regarding modification of these and other industrially important traits in barley.

12.4 Intellectual property and patenting issues in cereal biotechnology

Intellectual property protection is a strong incentive for innovation and development of new processes and products. Patent issues are very complex, particularly at the international level. Non-uniformity and the long time frames often required for a patent to issue from the time of filing, often measured in years, complicate the picture further. Coupled with the award of some very broad claims in key crops, the above shortcomings contribute to the general confusion. Often, claims and counterclaims made by competing organizations are the subject of bitter legal challenges and court disputes. It is not surprising, therefore, that the public and scientists alike find the issue of intellectual property protection highly confusing and contentious.

Numerous patents cover the methods for the creation of transgenic plants (including bombardment-based technology and the instruments used to insert DNA into plants), genetic constructs and their components, specific transformation procedures, selectable or screenable markers and common promoters, as well as novel germplasm generated as a result of a particular genetic modification. However, it is unlikely that any single individual or organization could gain complete control over a specific crop because of the interdependent nature of genetic engineering experiments. For example, the biolistic concept for introducing genes via particle bombardment has only been patented in the USA. Consequently, laboratories outside the United States are free to use this concept for the genetic engineering of crops which will not be imported into the USA. However, a number of instruments have been patented globally and this might present problems in the free use of the technology. Nevertheless, academic research is not threatened by patents since patents usually do not come into effect until the commercialization stage of product development.

The patent status of transgenic cereals is unclear. It appears that transgenic maize produced via particle bombardment will be the subject of a patent filed by DeKalb Genetics. It is not clear whether any other company or organization will have control through patents covering other cereals because one of the key criteria for having a patent granted (novelty) is in question once a specific technology has been shown to work for a few crops (the other two criteria are usefulness and reproducibility). An additional level of patent coverage will be provided by patents with much narrower claims, often focusing on specific genes or particular processes. Examples which fall into this category include technology for generating male sterility for hybrid production, developed by Plant Genetic Systems, Inc. (Belgium) and biopesticides based on *Bacillus thuringiensis* (Bt) toxins (described in more detail later in this chapter). A legal battle is underway at present between Mycogen and the Monsanto company for control of this key technology (Anon., 1996). Mycogen and Ciba-Geigy are marketing transgenic maize seed engineered with Bt-based resistance to the European corn borer. Monsanto recently won a court decision on a patent suit Mycogen had filed against Monsanto for its alleged infringement of Mycogen's US patent covering synthesizing Bt genes to resemble plant genes to enhance Bt toxin expression in transgenic plants. The court decided that Monsanto had utilized the technology prior to the patent's issue date and was exempt from infringement. Mycogen is considering appealing the decision. This is an extremely complex situation as issues of process versus composition come into play and it is very difficult, even for legal experts, to decide the outcome of such a dispute.

In the case of maize, the company or organization which gains control over the crop will itself fall under a more general patent which covers particle bombardment as well as a second patent covering the instrumentation for DNA delivery. It is obvious therefore, that at least in the commercial sector, there has to be extensive cross-licensing in order to allow introduction of products into the market. Another way of securing patent protection is for a larger company to acquire a patent-holding company. This is exemplified by the acquisition of a large part of DeKalb by Monsanto. This allows Monsanto access to the key patent for maize, assuming this holds against legal challenges in the courts. It is not clear how other companies will deal with such complex situations.

No other patents have been issued at present covering other cereals, however, it is very likely that such patents have been filed and will become public in due course. It is likely that specific patents providing protection for engineered cereals (and also other crops) will be narrower in scope and will most likely protect germplasm created with specific genes for particular applications or uses.

12.5 Health, safety and environmental issues

In terms of development of pest resistance strategies, particularly those involving single genes, it is important to ensure that genetic engineering is a component of a well thought-out integrated pest management system to minimize the probability of pests developing resistance to the specific gene product, thus rendering it useless. A classical example is the use of the insecticidal crystal protein from *B. thuringiensis*. It is now broadly recognized that insects are capable of developing resistance to any chemical control agent, including Bt. Consequently, in the USA, the Environmental Protection Agency requires the submission of a multicomponent pest resistance strategy as part of its approval process. It is also important to consider ecological ramifications of introducing new genes, particularly if the gene product is not specific for a particular class of pests. Herbicide resistance is also an issue as there might be the possibility of horizontal gene transfer to closely related weeds, in specific cases. However, this aspect has perhaps been overemphasized, particularly in Europe.

In the case of experiments in which resistances of plants to viruses are targeted, it is important to consider carefully whether there is any likelihood of generating new viral strains as a result of accidental release of genes which may recombine with other viruses to create novel strains with unpredictable results.

12.6 Legislative and labelling position

In the United States, major transgenic crops, including cereals such as maize, have been deregulated for specific traits. Thus, it is no longer necessary to seek permission to field-test insect and herbicide resistant maize. All that is required is notification of the intent to carry out such trials. Genetically engineered maize expressing insect and herbicide resistance is now on the market in the USA. It remains to be seen how soon transgenic crops, including cereals, will be introduced into the market in Europe. It is clear that the legislative and labelling position is very different in Europe but it is not clear whether all the reasons for this difference are based on sound scientific arguments or whether other factors come into play. It is very likely, however, that transgenic crops, including cereals, will eventually reach the market in Europe, particularly if a clear benefit to the user or consumer can be demonstrated. Although transgenic cereals modified for their agronomic traits are not currently labelled, the situation may change in Europe depending on how the new European Regulation on Novel Foods, enacted in May 1997, is eventually implemented (see Chapter 4 for further details).

12.7 Consumer acceptance and marketing

Products of plant engineering are rapidly approaching commercialization. A number of engineered crops, including the first transgenic cereals, are now in the market. The first cereal crop to be commercialized is maize engineered for insect and herbicide tolerance. We can expect to see additional traits introduced into maize, wheat and rice to reach the market in the late 1990s.

The USA has taken the lead in the commercialization of products of genetic engineering and indeed there now exists a climate which is conducive to deregulation of a number of engineered crops. A number of companies have petitioned successfully for the unrestricted use of transgenic crops, including cereals such as maize, engineered with the Bt gene for resistance against the European corn borer (Ciba-Geigy and others) as well as herbicide resistance (glufosinate and glyphosate, AgrEvo and Monsanto, respectively). The climate in Europe, however, is very different. A number of active environmental groups and non-government organizations have been successful in introducing restrictive guidelines for the release of transgenic crops. These groups have also influenced the general public and there are indeed various levels of scepticism and concern amongst different European countries as to the safety and acceptability of products of recombinant DNA technology. Thus, a clear dichotomy exists between consumer acceptance in the USA and in Europe.

The European Union has recently blocked the approval of genetically engineered maize designed to resist the European corn borer, which devastates maize all over the world (Coghlan, 1996). These plants were also engineered with a herbicide resistance gene. The plant has already been approved for sale in the USA and Canada. France had initially approved the transgenic maize plant, but Britain, Sweden, Austria and Denmark voted against approval, with four other countries (Germany, Italy, Luxembourg and Greece) abstaining. Decision now rests with the Council of Ministers.

It is interesting to examine the very different reasons for rejection of the GM maize in various European countries. Britain objected due to the presence of an antibiotic resistance gene in the transgenic plants, which precipitated fears that this gene might be transferred to bacteria thus making the therapeutic use of the antibiotic ineffective. Sweden, Austria and Denmark objected on the grounds that the maize plants should be labelled as genetically engineered. They also expressed concern that insects might develop resistance to the Bt toxin, and that the gene for herbicide resistance might spread to weeds. It is beyond the scope of this chapter to argue for or against these views. It is interesting, however, to note the mixed response that this decision precipitated amongst various groups in Europe. Organizations such as Greenpeace saluted the rejection, characterizing it as a step in the right direction. European biotechnology companies were alarmed, arguing strongly that Europe's strict laws, sometimes not based on rational scientific facts, will handicap European industry, while rivals in the USA and Japan enjoy less restrictive regulations (Coghlan, 1996).

12.8 Future prospects

Future directions in the genetic engineering of important cereals follow those for other crops. Agronomic characteristics include mostly single gene traits such as control of weeds, resistance to insect pests and diseases caused by bacteria, fungi or viruses. More complex agronomic characteristics which are controlled by multi-gene families include tolerance to biotic stresses such as heat, cold, drought, salt, etc. and also tolerance to abiotic stresses such as heavy metals, etc. Food processing traits include enhanced nutritional quality of food crops, delayed ripening of fruits, modifications in colour, flavour, texture, as well as modification of oil and starch composition. Safety issues can also be addressed by eliminating toxic or antinutritional factors from food products. Finally, industrial uses include the creation of modified and speciality oils, alternative uses of crops to include production of industrial and speciality enzymes and proteins, and also the production of recombinant macromolecules such as antibodies and vaccines for human and animal healthcare. These targets need to be selected carefully, however, to ensure that the specific macromolecules are produced economically and more efficiently in plants as compared to microorganisms, mammalian cultures or transgenic animals. An additional use is in bioremediation in which proteins capable of binding heavy toxic metals might be engineered into crops for cleaning up contaminated sites.

References

Anon. (1985) The Rockefeller Foundation solicits research proposals on the genetic engineering of rice. *Plant Mol. Biol. Reporter.* 3: 145–6.

Anon. (1996) Mycogen–Monsanto Bt patent battle; Round Two. *Biotech Reporter*, April issue. p. 4.

Armstrong, C. *et al.* (1995) Field evaluation of European corn borer control in progeny of 173 transgenic corn events expressing an insecticidal protein from *Bacillus thuringiensis. Crop Sci.* 35: 550–7.

Becker, D., Brettschneider, R. and Lorz, H. (1994) Fertile transgenic wheat from microprojectile bombardment of scutellar tissue. *The Plant Journal*, 5, 299–307.

Bittel, D. C. *et al.*, (1996) Lysine accumulation in maize cell cultures engineered with a lysine-insensitive form of maize dihydrodipicolinate synthase. *Theor. Appl. Genet.*, 92: 70–77.

Cao, J. *et al.* (1992) Regeneration of herbicide resistant transgenic rice plants following microprojectile-mediated transformation of suspension culture cells. *Plant Cell Reports*. 11: 586–91.
Christensen, A. H. and Quail, P. H. (1996) Ubiquitin promoter based vectors for high level expression of selectable and/or screenable marker genes in monocotyledonous plants. *Transg. Res.*, 5: 213–18.
Christou, P., Ford, T. L. and Kofron, M. (1992) The development of a variety-independent gene-transfer method for rice. *Trends in Biotech.*, 10: 239–46.
Christou, P. (1993) Particle gun mediated transformation. *Current Opinion in Biotech*, 4: 135–41.
Christou, P. and Ford, T. (1995) The impact of selection parameters on the phenotype and genotype of transgenic rice callus and plants. *Transgenic Research*. 4: 44–51.
Coghlan, A. (1996) Europe halts march of supermaize. *New Scientist*. May 4. p. 7.
D'Halluin, K. *et al.* (1992) Transgenic maize plants by tissue electroporation. *The Plant Cell*, 4, 1495–505.
Datta, S. K. *et al.* (1990) Genetically engineered fertile Indica-Rice recovered from protoplasts. *Bio/Technlogy*. 8: 736–40.
Datta, S. K. *et al.* (1992) Herbicide-resistant Indica rice plants from IRRI breeding line IR72 and PEG-mediated transformation of protoplasts. *Plant Mol. Biol.*, 20: 619–29.
Fromm, M. E. *et al.* (1990) Inheritance and expression of chimeric genes in the progeny of transgenic maize plants. *Bio/Technology*, 8, 833–9.
Fujimoto, H. *et al.* (1993) Insect resistant rice generated by introduction of a modified d-endotoxin gene of *Bacillus thuringiensis*. *Bio/Technology*, 11: 1151–5.
Funatsuki, H. *et al.* (1995) Fertile transgenic barley generated by direct DNA transfer to protoplasts. *Theor. Appl. Genet.*, 91: 707–12.
Gordon-Kamm, W. J. *et al.* (1990) Transformation of maize cells and regeneration of fertile transgenic plants. *The Plant Cell*, 2, 603–18.
Gordon-Kamm, W. J. *et al.* (1991) Transformation of maize using microprojectile bombardment: an update and perspective. *Vitro Cell Dev. Biol.*, 27, 21–27.
Grimsley, N. *et al.* (1987) Agrobacterium-mediated delivery of infectious maize streak virus into maize plants. *Nature*, 325, 177–9.
Hagio, T. *et al.* (1995) Production of fertile transgenic barley (*Hordeum vulgare* L.) plant using the hygromycin-resistance marker. *Plant Cell Reports*. 14: 329–34.
Hayakawa, T. *et al.* (1992) Genetically engineered rice resistant to rice stripe virus, an insect-transmitted virus. *Proc. Natl. Acad. Sci.*, USA, 89: 9865–9.
He, D. G. *et al.* (1994) Transformation of wheat (*Triticum aestivum* L.) through electroporation of protoplasts. *Plant Cell Reports*.
Heng, Z. *et al.* (1992) *In vitro* morphogenesis of corn (*Zea mays* L.). 1. Differentiation of multiple shoot clumps and somatic embryos from shoot tips. *Planta*. 187: 483–9.
Hiei, Y. *et al.* (1994) Efficient transformation of rice (*Oryza sativa* L.) mediated by *Agrobacterium* and sequence analysis of the boundaries of the T-DNA. *Plant J.*, 6: 271–82.
Irie, K. *et al.* (1996) Transgenic rice established to express corn cystatin exhibits strong inhibitory activity against insect gut proteinases. *Plant Mol. Biol.*, 30: 149–57.
Ishida, Y. *et al.* (1996) High efficiency transformation of maize (*Zea mays* L.) mediated by *Agrobacterium tumefaciens*. *Nature Biotechnology*. 14: 745–50.
Jahne, A. *et al.* (1994) Regeneration of transgenic microspore-derived, fertile barley. *Theor. Appl. Genet.*, 89: 525–33.
Keil, M. *et al.* (1986) Primary structure of a proteinase inhibitor II gene from potato (*Solanum tuberosum*). *Nucl. Acids Res.*, 14: 5641–50.
Klein, T. *et al.* (1987) High velocity microprojectiles for delivering nucleic acids into living cells. *Nature*, 327: 70–73.
Koziel, M. G. *et al.* (1993) Field performance of elite transgenic maize plants expressing an insecticidal protein derived from *Bacillus thuringiensis*. *Bio-Technology*, 11, 194–200.
Matsuoka, M. *et al.* (1993) Expression of a rice homeobox gene causes altered morphology of transgenic plants. *The Plant Cell*. 5: 1039–48.
Nehra, N. S. *et al.* (1994) Self-fertile transgenic wheat plants regenerated from isolated scutellar tissues following microprojectile bombardment with two distinct gene constructs. *The Plant Journal*, 5, 285–97.

Oard, J. H. *et al.* (1996) Development, field evaluation and agronomic performance of transgenic herbicide resistant rice. *J. Molec. Breeding*, in press.
Rhodes, C. A. *et al.* (1988) Genetically transformed maize plants from protoplasts. *Science*, 240, 204–207.
Ritala, A. *et al.* (1994) Fertile transgenic barley by particle bombardment of immature embryos. *Plant Molecular Biology*, 24, 317–25.
Shewry, P. R. *et al.* (1995) Biotechnology of Breadmaking: Unravelling and manipulating the multi-protein gluten complex. *Bio/Technology*, 13: 1185–90.
Shimada, H. *et al.* (1993) Antisense regulation of the rice *waxy* gene expression using a PCR-amplified fragment of the rice genome reduces the amylose content in grain starch. *Theor. Appl. Genet.*, 86: 665–72.
Sivamani, E. *et al.* (1996) Selection of large quantities of embryogenic calli from indica rice seeds for production of fertile transgenic plants using the biolistic method. *Plant Cell Reports*. 15: 322–237.
Song, W. Y. *et al.* (1995) A receptor kinase-like protein encoded by the rice disease resistance gene Xa21. *Science*. 270: 1804–6.
Toenniessen, G. H. (1995) Plant biotechnology and developing countries. *Trends in Biotechnology*. 13: 404–9.
Toriyama, K. *et al.* (1988) Transgenic rice plants after direct gene transfer into protoplasts. *Bio/Technology*. 6: 1072–4.
Uchimiya, H. *et al.* (1993) Bialaphos treatment of transgenic rice plants expressing a bar gene prevents infection by the sheath blight pathogen (*Rhizoctonia solani*). *Bio/Technology*, 11: 835–6.
Vasil, V. *et al.* (1992) Herbicide resistant fertile transgenic wheat plants obtained by microprojectile bombardment of regenerable embryogenic callus. *Bio/Technology*. 10, 667–74.
Vasil, V. *et al.* (1993) Rapid production of transgenic wheat plants by direct bombardment of cultured immature embryos. *Bio/Technology*, 11, 1553–8.
Wan, Y. and Lemaux, P. G. (1994) Generation of large numbers of independently transformed fertile barley plants. *Plant Physiol.*, 104: 37–48.
Wan, Y., Widholm, J. M. and Lemaux, P. G. (1995) Type I callus as a bombardment target for generating fertile transgenic maize (*Zea mays* L.). *Planta*, 196: 7–14.
Weeks, J. T., Anderson, O. D. and Blechl, A. E. (1993) Rapid production of multiple independent lines of fertile transgenic wheat (*Triticum aestivum*). *Plant Physiol.*, 102, 1077–84.
Wunn, J. *et al.* (1996) Transgenic indica rice breeding line IR58 expressing a synthetic cryIA(b) gene from *Bacillus thuringiensis* provides effective insect pest control. *Bio/Technology*, 14: 171–6.
Xu, Y. *et al.* (1996) Regulation, expression and function of a new basic chitinase gene in rice (*Oryza sativa* L.). *Plant Mol. Biol.*, 30: 387–401.
Yang, N. S. and Christou, P. (1994) *Particle Bombardment Technology for Gene Transfer*. Oxford University Press, New York, Oxford.
Zhang, S. *et al.* (1996) Regeneration of fertile transgenic indica (group 1) plants following microprojectile transformation of embryogenic suspension culture cells. *Plant Cell Reports*. 15: 465–9.
Zhang, W. and Wu, R. (1988) Efficient regeneration of transgenic plants from rice protoplasts and correctly regulated expression of the foreign gene in the plants. *Theor. Appl. Genet.*, 6: 835–40.
Zheng, Z. W. *et al.* (1995) The bean seed storage protein beta-phaseolin is synthesized, processed and accumulated in the vacuolar type II protein bodies of transgenic rice endosperm. *Plant Physiol.*, 109: 777–86.
Zhong, H. *et al.* (1996) The competence of maize shoot meristems for integrative transformation and inherited expression of transgenes. *Plant Physiol.*, 110: 1097–107.
Zhou, H. *et al.* (1995) Glyphosate-tolerant CP4 and GOX genes as a selectable marker in wheat transformation. *Plant Cell Reports*. Vol. 15: 159–63.

Glossary

ADH promoter	Promoter of the alcohol dehydrogenase gene. Often used to achieve over-expression of genes in yeast.
Active oxygen method (AOM)	Method used to evaluate the stability of vegetable oils.
Agrobacterium	A common soil bacterium that can naturally infect plant cells with DNA. Often used as a tool for gene transfer in plant biotechnology.
amdS gene	A fungal gene which, if introduced into yeast, confers ability to grow on acetamide as sole nitrogen source. Used as selection marker.
Antisense molecule	A chemical derivative of DNA or RNA which binds specifically to a (sense) strand of RNA or to a DNA helix and stops it from encoding a protein.
Bacteriocin	A small protein or peptide with antimicrobial properties often produced by lactic acid bacteria.
Bar	A gene coding for phosphinothricin acetyltranferase, an enzyme which detoxifies the active ingredient in gluphosinate-based herbicides (e.g. Basta or Ignite).
Bases	There are four DNA bases: adenine, thymine, guanine and cytosine. The order of the bases provides the code in genetic information.
Basta	Trade name for a non-selective herbicide containing gluphosinate as the active ingredient.
Bialaphos	Non-selective herbicide.
Biolistics	Acceleration of metal particles coated with genetic material (DNA or RNA) into living cells using gunpowder. Also known as particle bombardment.
Callus	Dedifferentiated plant tissue *in vitro*.
cDNA	Complementary DNA derived from messenger RNA enzymatically *in vitro* before insertion into a vector.

Chromosome	A string of many genes. Each chromosome is a densely coiled molecule of DNA and looks like a tiny thread when observed under a microscope. The number of chromosomes varies from species to species. Humans have a set of 46 chromosomes in every cell.
Clone	One or more genetically identical organisms. Bacteria clone naturally. Frogs and sheep have been cloned artificially.
Cloning	Producing large numbers of identical cells or organisms from a single ancestor. Genes transferred into a single cell which is then allowed to grow to produce many identical cells are often described as having been cloned.
Conjugation	The natural exchange of genetic material between organisms such as the lactic acid bacteria.
Conjugative plasmid	A plasmid capable of directing its own transfer between donor and recipient strains.
Construct	A gene construct is a sequence of DNA artificially synthesised by genetic engineering.
Cryptic plasmid	A plasmid with no known function.
Cystatin	A gene encoding for a protease inhibitor active against parasitic worms.
DNA	Deoxyribonucleic acid. The genetic material of a cell is composed of DNA. It consists of two chains of bases joined together as a double helix.
DNA probe	A small piece of labelled DNA that can bind to a particular DNA sequence, e.g. a gene or part of a gene.
Electroporation	Physical method for forcefully introducing DNA into plant or microbial cells using an electric pulse.
Embryogenic	Tissue capable of regenerating through somatic embryogenesis.
Explants	Starting material often used in plant biotechnology for tissue culture and transformation experiments.
Expression	Appearance of a particular trait or active production by a cell of a protein determined by a gene.

FBP1 gene	Gene encoding the gluconeogenic enzyme fructose-1,6-biphosphatase. Used together with PCK1 gene to introduce ATP-spoiling reactions in yeast, leading to increased gas production under stress conditions.
FAD gene	Gene encoding a fatty acid desaturase enzyme.
Fatty acid desaturase	Enzyme which catalyses the introduction of a double bond into a fatty acid.
Gene	The smallest segment of DNA containing a hereditary message. Each gene is a basic unit of heredity, passing traits from one generation, onto the next.
Genetic code	The sequence of nucleotides that determines protein synthesis in cells.
Genetic modification	Term to describe a series of techniques used to transfer genes from one organism to another or to alter the expression of an organism's genes. Also referred to as genetic manipulation or genetic engineering.
Genotype	The genetic make-up of an individual organism.
Germline	Germline cells are or eventually produce reproductive cells which carry the entire genetic information passed on to the next generation, e.g. egg and sperm in humans.
Gluphosinate	Active ingredient in non-selective herbicides such as Basta and Ignite.
GMO	Genetically modified organism. An organism or its progeny into which DNA had been transferred by gene technology. Also called a transgenic organism.
gusA	A gene from *Escherichia coli* encoding the enzyme β-glucuronidase. Often used as a selection marker.
G418 gene	Gene encoding a protein that confers resistance to phleomycin. Often used as a selection marker in yeast.
Heterologous	Recombination of DNA sequences originating from different species.
Homologous	Recombination of DNA sequences originating from different strains of the same species.

Ignite	Trade name for a non-selective herbicide containing gluphosinate as the active ingredient.
Integration	The process of inserting a piece of DNA into the chromosome of a cell rather than leaving it as a plasmid independent of the chromosome.
Intron	Region within a gene that does not code for a protein sequence and is separated by coding regions known as exons. During transcription (formation of messenger RNA from DNA), the introns are excised and the exons are spliced (joined) together.
MAL genes	Genes encoding enzymes involved in the metabolism of maltose, the main carbon source for yeast in bread dough.
Marker gene	A gene which is used to differentiate easily between organisms which have been successfully genetically modified from those which have not, e.g. by resistance to an antibiotic. Markers are used extensively as reference points to map other genes.
Microspores	Pollen progenitor cells.
mRNA	Messenger ribonucleic acid carries information from the genes to the protein-synthesising machinery of the cell.
Multi-copy plasmid	A plasmid which can exist within a cell as a number of identical copies.
Mutation	A random change in genetic material that can change a host of functions and may be lethal.
NTH gene	Gene encoding enzymes involved in the breakdown of trehalose.
Nucleotide	The building block of the DNA molecule consisting of an organic base, a phosphate and a sugar.
Oxidative stability index (OSI)	An automated method for the determination of the stability of vegetable oils.
Particle bombardment	Technique for accelerating metal particles coated with genetic material into living cells using gunpowder. Also known as biolistics.

Phenotype	The outward appearance and behaviour of an organism resulting from the interaction between its genetic constitution and the environment.
Phytocystatin	A plant cystatin (see cystatin above).
Plaque	Zone of lysis (clearing) seen in a normally opaque bacterial cell lawn as a consequence of an attack initiated by a single bacteriophage (bacterial virus).
Plasmid	A small self-replicating ring of DNA found in many yeasts and bacteria. Plasmids are widely used in genetic modification because they are able to pass easily from one cell to another.
PCK1 gene	Gene encoding the gluconeogenic enzyme phosphoenolpyruvate carboxykinase. Used together with FBP1 gene to introduce ATP-spoiling reactions in yeast, leading to increased gas production under stress conditions.
PCR	Polymerase chain reaction. A method for repeatedly duplicating very small amounts of DNA in order to have sufficient amounts for analysis. Used in diagnosis of diseases and in forensic medicine.
Promoter	A piece of DNA (a DNA sequence) ahead of the main coding region of a gene which controls the expression of that DNA into protein.
Protoplast	Plant or microbial cell with walls removed using enzyme digestion.
Radioallergosorbent test (RAST)	Method used to test the allergenicity of a food or food component.
Rare mating	A procedure for producing hybrid strains of yeast from organisms which are very reluctant to mate. Many millions of cells are mixed together and a suitable strong selection procedure used to isolate the hybrids made by the very few cells which do mate.
Recombinant DNA	The hybrid DNA produced by joining pieces of DNA from different sources.
Replicon	Any circular piece of DNA which can replicate independently.
Restriction enzyme	An enzyme that cuts DNA molecules at specific points depending upon the sequence of base pairs.

Reverse transcriptase	An enzyme that can synthesise a single strand of DNA (complementary DNA or cDNA) from messenger RNA. This is the reverse of the normal direction in which genetic information is processed i.e. DNA→RNA→protein.
RNA	Ribonucleic acid. Plays an intermediary role between DNA and proteins.
Roundup	A non-selective herbicide.
Selective marker	A gene included in a plasmid to enable a simple procedure to be used to isolate the cells which have undergone transformation in genetic modification procedures.
Self cloning	The removal of DNA from a cell followed by re-insertion of all or part of that DNA (with or without further enzymic, chemical or mechanical steps) into the same cell or into closely related strains of the same species.
Sense suppression	A method of gene silencing.
Scutellum	Modified cotyledon of monocots; often the preferred tissue for transformation experiments.
Solid fat content (SFC)	The ratio of solid to liquid phases in an oil or fat at any given temperature.
Spheroplast fusion	A procedure for producing hybrid strains of yeast from organisms which are very reluctant to mate. The two parent cells are treated with an enzyme to remove their cell walls (producing spheroplasts). On mixing together the cells fuse after which the cell wall of the new hybrid is allowed to regrow.
SIT genes	Non-essential baker's yeast genes used as a site for the introduction of modified genes.
Terminator	A piece of DNA at the end of the main coding region of a gene which signals the end of that gene.
trans fatty acid	Unsaturated fatty acid in which one or more of the double bonds has hydrogen atoms in the *trans* (on opposite sides of the bond) rather than in the *cis* (on the same side of the bond) configuration.
Transcription	Formation of messenger RNA from the DNA template.

Transformation	The process of changing the properties of a cell by inserting new genes during genetic modification procedures.
Translation	The formation of protein from messenger RNA.
Transgenic	An organism that has been genetically modified to contain DNA from another organism.
Transposon	A small segment of DNA that can move from one location to another within a DNA strand. A gene which is carried by a transposon is often referred to as a 'jumping gene'.
Triacylglycerol	Molecule containing three fatty acids attached to a glycerol backbone. Main molecular constituent of oils and fats.
2μ plasmid	A multicopy plasmid in yeast.
Vector	The vehicle by means of which DNA fragments can be introduced and replicated into a host organism. The most common vectors are those based on plasmids.
YeP	Yeast episomal plasmid.
YiP	Yeast integration plasmid.
Zygotic	Sexually-derived, as opposed to somatic, tissue.

Index